DEWALT®

PLUMBING LICENSING

EXAM GUIDE

American Contractors Exam Services

Created exclusively
for DEWALT by:

PAL publications®

www.dewalt.com/guides

ISBN 0-9770003-6-2

Pal Publications, Inc.
800 Heritage Drive
Suite 810
Pottstown, PA 19464

First edition published 2007

Printed in the United States of America

Titles Available From DEWALT:

DEWALT Trade Reference Series

Blueprint Reading Professional Reference
Construction Professional Reference
Construction Estimating Professional Reference
Datacom Professional Reference
Electric Motor Professional Reference
Electrical Estimating Professional Reference
Electrical Professional Reference
HVAC/R Professional Reference – Master Edition
Lighting & Maintenance Professional Reference
Plumbing Professional Reference
Residential Remodeling & Repair Professional Reference
Security, Sound & Video Professional Reference
Spanish/English Construction Dictionary – Illustrated Edition
Wiring Diagrams Professional Reference

DEWALT Exam and Certification Series

Building Contractor's Licensing Exam Guide
Electrical Licensing Exam Guide
HVAC Technician Certification Exam Guide

CONTENTS

DEWALT PLUMBING LICENSING EXAM GUIDE

by
American Contractors Exam Services

Notes

ABOUT THIS BOOK

Congratulations! If you are reading this book, it is likely you are pursuing a new opportunity. Whether you are preparing to take an exam in order to start your own plumbing business, to enhance the qualifications of an existing company, or to qualify the company for which you now work, you are taking deliberate measures to improve an existing situation. So sit back, relax, and rest assured, the time you spend with this book will significantly improve your chances of passing the exam you are about to take.

Time for Test Mode

Preparing for and taking an exam requires a specific mindset. It requires planning for the upcoming event and making a conscious effort to control your anxiety. For most, the mere thought of taking an exam will cause at least a feeling of butterflies churning in the stomach. Some may even react more severely. Don't fear, controlled test anxiety is good. It creates a sense of urgency to succeed. Identify your level of anxiety and control it.

How do you rate your level of anxiety? For some, perhaps the last test-taking ordeal was a high school final exam many years ago. The memory of this experience, coupled with the pressure of living up to the expectations of family, boss, or coworkers can be overwhelming. When children are aware that a parent will soon be taking an exam, they are likely to reciprocate the pressure applied to do well in school. Bosses often assume the exam will be a breeze for such a competent employee. Many times, a spouse's support is mistaken for unrealistic expectation. Peer pressure is not necessary. Do what you can to minimize it.

The pressures felt from the need to please and live up to the expectations of everyone around you can lead to undue stress and are not necessary. Simply keep the fact that you are taking the exam a secret! Maybe everyone notices your study efforts but there is no need to announce the test date. Make your plans by scheduling the exam, and if you must explain, simply say you have an appointment. If no one knows your plans, the only pressure you will feel to succeed is your own. This pressure is healthy and probably necessary. Rather than dealing with the pressure of others, imagine the feeling when you announce your accomplishment after the fact. Work toward this goal and visualize success. Think positive.

Another way to eliminate stress may be to think of your first test date as a practice run. The worst case scenario is that passing the exam will take two attempts. The simple fact of the matter: It is not the end of the world if you do not pass on the first attempt. Most states and municipalities allow you to take the exam as many times as necessary without penalty. Make sure you are not limited by the number of times you can take an exam before using this suggestion. (We will explain later how to check this.)

Next, avoid becoming irritated that the exam is required. Forget the fact that you know your trade and feel you should not be required to pass an exam to prove yourself. Look at the experience as an opportunity. The knowledge you gain from preparing for the exam will far outweigh the effort. Without a doubt, you will gain information that will improve your efficiency and trade expertise and could even save you money.

Remember the old carpenter's advice to measure twice and cut once? Similarly, when it comes to test-taking, it is critical to be precise, careful, and methodical as you prepare for exam day. Apply this philosophy as you take the practice exams in this book—read the question twice and choose the answer carefully. The exams are typically administered in a multiple-choice format and it is not uncommon for at least two of the choices to appear correct, depending on how the question is read. Remember, there is only one correct answer. One word in the question can change the entire meaning of the question. Be careful.

After you have switched to test mode, it is time to begin preparing for the exam.

Preparing for The Exam

The first step of preparation is to identify the exam required to accomplish your goal. If you are taking a state-required exam to become a licensed contractor, begin by visiting www.becomealicensedcontractor.com. Select your state, choose plumbing under the listing of test types, and the test requirements will be found. The site will likely lead you to the candidate bulletin provided by the approved testing company.

The candidate bulletin is the test-taker's rulebook. It provides specific information including the addresses of testing centers, identification requirements, security procedures, and anything else necessary for taking the exam. In preparing for the exam, it is important to identify the answers to the following questions using the candidate information bulletin:

1. **What are the approved references?** This is, by far, the most important piece of information to know prior to beginning your exam preparation mission. Most plumbing exams are based on

Notes

Notes

both the International Plumbing Code and the International Gas Code. In addition, many exams include content that will be found in an OSHA Standards Part1926 book, a plumbing theory book and a project management book. After identifying the approved reference materials, you can eliminate the study exams in this book that are not included in your exam. Do not spend valuable time studying information that is not included on your exam.

2. **Are the approved references allowed into the examination center?** In most cases the books are allowed in the examination room to be available to you throughout the exam. While it would seem to make the exam extremely easy, if you are not familiar with how to use the books, it might as well be a closed book exam.
3. **Is tabbing and highlighting of the books allowed?** Typically, if the approved references are allowed to be taken into the exam room, they are allowed to be highlighted and tabbed prior to test day. Many rules allow you to underline but prohibit you from making notes in the reference materials. The candidate bulletin will specifically call for permanent tabs. Permanent tabs are those which can not be easily removed. Post-It® tabs are generally not allowed. If any of the rules are broken, your reference materials can be banned from the test site.
4. **What is the time frame for taking the exam?** This is very important because you will want to simulate the time allowed as you work through the study exams throughout this book. Simply take the allotted time in minutes and divide by the number of questions on the exam to determine the time allotment per question. You can now easily multiply the number of questions in a particular section of this book by the average time allowed per question to set the time frame for working through the practice exams. In the beginning, you will want to allot additional time until you become familiar with the reference materials.
5. **What is the content outline of the exam?** It is important to pay attention to the breakdown of the exam for the purposes of allotting your study time. If only 5 out of 80 questions pertain to OSHA, you should spend less time studying this subject and concentrate more on the areas representing the largest portion of the exam.

USING THIS EXAM GUIDE

Before starting the exam preparation process using this guide, make sure you have highlighters and tabs readily available. Begin by identifying the table of contents, the index, and the glossary of each book. Place a tab on each of these sections for easy access. The table of contents will divide the book into chapters or subjects and will be used frequently. The index is an alphabetical listing of key words found throughout the book and should be your starting point for finding an answer to a specific question. If the answer is not found using the index, identify the chapter according to the subject of the question using the table of contents. The glossary is an alphabetized listing of terms and can serve as a useful source for answering questions quickly. A glossary may or may not be found in the approved references.

Each section of this exam guide is based either on a particular plumbing subject, general project management information, OSHA regulations, or a chapter of the *International Plumbing Code* or *International Gas Code* and is so identified. As you answer each of the questions from the study exam, highlight the answers. This will help you become familiar with the book and will strengthen your ability to quickly reference important code items. Remember, the purpose of each practice exam is not to test your trade knowledge, it is to provide an exercise of how to navigate and use the approved reference(s). If you feel you know the answer to the question, you should verify the answer using the applicable book. Pay very close attention to the tables in the references and the subject matter of each chapter.

When it comes to tabbing your reference materials, be careful not to overdo it. Placing too many tabs throughout your reference materials will serve more as a hindrance than a useful tool. Remember, when using the tabs, the more tabs you use, the longer it will take to read through each one. Using the index will likely save you time in your search for specific information. It is highly recommended that you tab the chapters, table of contents, index, and useful tables.

Going into the exam, make sure you are familiar with the subject matter of each book. It is imperative that you understand the layout and format of each of the references. This is accomplished through the process of answering the questions in this guide.

Prior to exam day, remove the index in any of the approved reference books bound with a three ring binder, staple it together and place it in the front pocket of the book. This will allow easy access to the index by being able to place it on the testing table beside the reference rather than requiring you to flip back and forth between the index and the body of the book.

On average, exams allow about 3 minutes per question. When you begin using the study exams in this book, allow yourself between 8

Notes

Notes

and 10 minutes per question. As you become more familiar in your use of the reference materials, decrease this time allotment to about 5 minutes. By the time you complete the final exams, allow yourself only the time allotted on your exam.

It is important to immediately devise a time management strategy that works for you. Your goal by test day is to have a plan of action on how to work through the exam. As you are preparing for the exam, keep in mind that the clock will be working against you. Decide the maximum amount of time you will spend on one question before moving on to the next one. You do not want to run out of time.

WHAT TO BRING TO THE EXAM

Prior to test day, make sure you get a good night's rest and arrive armed with the following items:

A copy of the candidate information bulletin. Many times the proctor of the exam is inexperienced. If you are told the tabs are not allowed, you need to be able to defend yourself by referring to the candidate bulletin (the rulebook).

Bottle of water. The clock does not stop during the exam. There are no "hold" buttons. Should you need a sip of water and have to run to the water fountain or bathroom, your timer will continue to count down. Although you have little choice when it comes to a bathroom break, at least be prepared for the dry mouth syndrome.

Magnifying glass. Many of the documents/diagrams used throughout the exam are difficult to read. Save the frustration and headache-causing eyestrain and use a magnifying glass.

Two pencils and a pencil sharpener. Arrive with at least two pencils and a pencil sharpener, especially if you are taking the exam using the old fashioned pencil and paper format. Tests are still administered this way in several states and municipalities.

Two calculators. Remember Murphy's law, if it can go wrong it probably will on test day. Have a contingency plan for everything. If you insist on using your favorite calculator and it happens to be one that you are not sure is allowed, such as a construction master, make sure you have a backup.

A great attitude. Make every attempt to remain calm, cool, and collected. This is easier to maintain if you have adequate rest the night before the exam. Do not cram and stay up until midnight. This will work against your tolerance level for aggravation.

Remember, frustration will only create tension and make everything more difficult.

Make sure you also understand the requirements for identification, payment methods, and proper exam registration documentation.

LET THE TEST BEGIN

The moment you sit in the "hot seat" to begin the exam, let your strategy unfold. Have a plan and stick with it. A few recommendations:

1. **Switch to Test Mode.** Many times the rules of thumb and the assumptions you make in the field will not work in the test world. While your background and experience can be an asset, don't allow them to get in the way on test day. Make no assumptions. If you are not 100 percent sure about an answer, verify it at some point. Deliberately focus and concentrate on each question.
2. **Arrange your work area neatly.** Stack your books to one side and if possible, stand them upright so they are each individually accessible. Decide that at least this one day you will be the most organized and careful person in the world.
3. **Download the memory.** That's right, download the information you are having trouble remembering. Transfer the formulas (and anything else you have been repeating since you walked into the test site) from your brain to the scratch paper provided by the proctor.
4. **Become familiar with the construction drawings and/or diagram booklet before you start the clock.** If your test is being administered by computer, you are in control of when the countdown begins. Take advantage of the control, but be careful not to push it. The proctor is only a few steps away and may prod you to begin the test if you wait too long.
5. **Answer the easy questions first.** Nothing will boost your confidence more than to run through a good portion of the exam answering questions based on information you recall from your studies. In contrast, your confidence level can diminish rapidly if you get distracted with a difficult question.
6. **"Mark" any question you answer and have doubt about.** If the test is computer based, you have the option to "mark" questions to review later. If you run out of time, the computer accepts the selected answer and does not penalize you for marking the question. By marking questions, if you have

Notes

Notes

additional time after answering all questions, you can verify the answers you selected.

7. **Leave the difficult questions unanswered and come back to these last.** As you skip questions, make a note on the scratch paper associating the question number with a specific book. This will allow you to categorize each of the unanswered questions by book saving valuable time and unnecessary frustration. After reading the last question on the exam, you will be allowed to revisit unanswered questions or marked questions. You can choose to go to the first of these in the group or to a specific question number.
8. **Do not run out of time.** Pay attention to the clock. Do not leave any question unanswered. Before you run out of time, select a choice for each question. Questions left unanswered will be counted against you. If you have to guess on a number of questions, improve your odds by selecting the same choice on each question. For example, select all As or all Cs but of course your goal is to manage your time and not have to guess on any of the questions.

The Dreaded Math

When it comes to the math portions of the exam, there are several very important and critical rules. It is in your best interest to convince yourself that for this one day, TEST DAY, you are going to take a methodical approach and deliberately follow each of the rules we are about to discuss.

1. **Always use a calculator.** This is probably the most critical and most important rule when it comes to math on the exam. Never perform mental calculations. Once you arrive at an answer, do the calculation again to check your answer. Repeat your calculations until you are certain the answer is correct.
2. **Use scratch paper to write down each step of math.** Professionals who use math on a daily basis, accountants, engineers, and the like, are sometimes the most difficult math students. Why? Because they refuse to write out the problem and their calculations, and they are reluctant to use a calculator for what they believe to be simple math. You do not want to fail this exam by one or two points; it hurts more to make a 69 than it does to make a 30. The point is, don't get in a hurry and miscalculate. Always write out your calculations and always calculate, at least twice, with a calculator.

3. **Never round until you get to the end.** This simply means that you should never round any of your numbers until you solve the equation. When necessary, round based on the choices or the directions provided in the question.

4. **Draw it out.** If you have to sketch apples and stick people, do it. Remember, you do not want to fail this test by one question. It is critical that you use your scratch paper to draw as you solve the equation to answer the question.

5. **Always multiply like numbers.** In other words, do not mix feet and inches. If you multiply 10 feet by 8 feet 6 inches, the answer is not 86, it is 85. You cannot simply multiply 10 by 8.6—you must first convert the 6 inches to feet.

6. **When converting inches to feet, round to the nearest hundredth of a foot.** Yes, this is an exception to rules. You round to the nearest hundredth when converting inches to feet because it puts your answer in the correct range of the choices provided and allows you to write it down more easily. You would not want to write 0.3333333333333 it would be much simpler to write 0.33.

Try these rules as you delve into the next section of the book, which covers the basics for performing necessary math calculations found on most plumbing exams. Good luck.

Notes

Notes

PART ONE
Math

The following section covers the basic math concepts you are likely to encounter on your exam. It begins with a review of the terminology and simple concepts necessary to perform the calculations. Following the review, the 25 questions will test your general understanding of math as it applies to the plumbing industry.

Notes

NUMBERS

Whole Numbers

Whole numbers are positive numbers without any fractions or decimals.

Example: $100.00; 8 hours; 12 wrenches

When we speak or write whole numbers, we automatically calculate certain numbers in our minds. . .

625: Six hundred plus twenty plus five
625: When we speak, we leave out the plus.
Six hundred twenty five

Mixed Numbers

Mixed number are whole numbers and parts of whole numbers. They can be expressed as fractions or as decimals.

Examples: $100.50; 7.5 hours; $5\frac{1}{3}$ inches

Decimal Numbers

A decimal number refers to part of a whole number. It is based on 10 or tenths broken down into hundredths, thousandths, etc., depending on how many numbers follow the decimal point.

Examples: 1.5 inches; 7.75 feet; 6.25 pounds

Fractional Numbers

A fractional number refers to part of a whole number expressed in a fraction instead of decimal form. It is possible to convert decimals to fractions and fractions to decimals.

Examples: $2\frac{2}{3}$ inches; $4\frac{1}{2}$" pipe; $1\frac{1}{4}$ miles

$$\frac{a}{b} = \frac{\text{numerator}}{\text{denominator}} = \frac{\text{part of a whole number}}{\text{the whole number}}$$

Fraction Bar

The Fraction Bar is the slash between the numerator and the denominator. It means to divide.

Example: $\frac{3}{4}$ means 3 divided by 4 or $3 \div 4$

Labeling Numbers

Labeling numbers is very important. On construction sites an estimator makes the calculations and architects draft the construction drawings. Labels are important to identify the unit of measurement. Label all calculations or lengths as inches, feet, yards, lengths of pipe, volume, pounds, etc.

Steps required for solving the equation:

1. Examine the problem and decide what information is given and what calculations are required.
2. Write down the information given and what requires an answer.
3. Decide which type of mathematical calculation is required to solve the problem.
4. Execute the complete operation.
5. Label the answer with the appropriate unit inches, feet, BTUs, etc.

MATHEMATICAL SYMBOLS

Symbol	Meaning
/	Fraction bar
$+$	Add
$-$	Subtract
$\times$	Multiply
$\div$	Divide
$=$	Equals
$\neq$	Does not equal
$\sqrt{}$	Square root
π	Pi = 3.14
( )	Perform the calculation in the parentheses first

PIPING CALCULATIONS

Plumbers will sometime or another calculate piping offsets. Offsets are pipe assemblies connecting a section of pipe from one line to another. The pipes are usually parallel. When using the right triangle theory, the length of the sides of a pipe offset can be calculated knowing the angle of the fittings and the length of one of the sides.

Notes

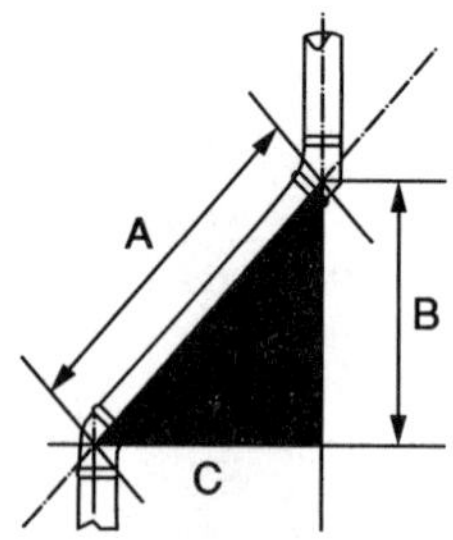

A. Hypotenuse = Travel
B. Altitude = Advance
C. Base = Offset

Three important lengths include:

A. **Travel** (Hypotenuse) which is the distance from the initial offset to the second fitting.

B. **Advance** (Altitude) is the distance parallel to the pipes between the offset fittings.

C. **Offset** (Base) is the distance the pipe is offset from its original position to the new position.

Some of the pipe fitting angles that are used by plumbers for pipe offsets are:

- 22½ degree elbow or 1/16 bend.
- 45 degree elbow or ⅛ bend.
- 60 degree elbow or ⅙ bend.
- 72 degree elbow or ⅕ bend.
- 90 degree elbow or ¼ bend.

The plumber usually calculates the length of the travel. To calculate the travel, the angle of the fittings and the length of the advance or offset is required. The known values are multiplied by a constant. (See the single pipe offsets constants table below.) A 45° elbow (⅛ bend) is widely used by plumbers. The offset constant for this is 1.414 and should be memorized.

TABLE—SINGLE PIPE OFFSETS CONSTANTS

Fitting	Fitting Angle	Travel = Offset ×	Travel = Advance ×	Advance = Offset ×	Advance = Travel ×	Offset = Advance ×	Offset = Travel ×
1/32	11¼°	5.126	1.019	5.027	0.981	0.199	0.195
1/16	22 1/1°	2.613	1.082	2.414	0.924	0.414	0.383
⅛	45°	1.414	1.414	1.000	0.707	1.000	0.707
⅙	60°	1.155	2.000	0.577	0.500	1.732	0.866
⅕	72°	1.051	3.236	0.325	0.309	3.078	0.951

Calculating Travel—Travel = Offset × Constant or Travel = Advance × Constant

Example:

Calculate the travel of a 45° elbow (⅛ bend) with an offset of 29".

1. Place the known values.
   Travel = Offset × 1.414 (taken from offsets table on page 14)
   Travel = 29" × 1.414 (offset is given value)

2. Calculate the travel.
   29" × 1.414 = 41.006"
   Travel = 41.006 approximately 41"

Calculate End-To-End

41 inches is the distance of the travel from center-to-center (C–C) measurement. Subtract twice the end-to center calculating the end-to-end (E–E) measurement. This will be the length of pipe you should cut. Rule-of-thumb is that you subtract twice the size of the pipe being cut to allow for the fitting on each end.

Example:

In the above calculation the C–C measurement was 41 inches. If the pipe offset was 2 inches the E–E measurement would be 37 inches.

41" − 4" (2" for both fittings) = 37"

Multiple Pipe Offsets

Installing multiple pipes require increased design ability by the plumber. Multiple pipes installed parallel to one another must be placed with a uniform spread. Spread is the center-to-center distance between the pipes. Offset elbows are installed evenly, guaranteeing an equal spread ensuring quality craftsmanship. 45 degree offsets are the best to use.

To install multiple pipes you must calculate the difference in the lengths of the pipes. The difference is added to or subtracted from the various pipe lengths used for the multiple pipe installation. Just like single pipe installations a constant value is used in the calculation.

Notes

MULTIPLE PIPE OFFSETS
Uniform Spread Constants
S = Spread d = difference pipe length

Fitting	Fitting Angle	Difference in length d = S ×	Travel = Offset ×
1/32	11¼°	0.099	5.126
1/16	22½°	0.199	2.613
⅛	45°	0.414	1.414
¼	60°	0.577	1.155
⅕	72°	0.727	1.051

Notes

Calculating Multiple Pipe Offsets

An equation is used to calculate multiple pipe offsets. Two of the three values are given and you calculate the third. To calculate multiple pipe offsets use the following equation:

$$d = S \times \text{constant}$$

The d value or difference equals the difference in length of the pipes. The value is obtained from the table on previous page and is based on the fitting size.

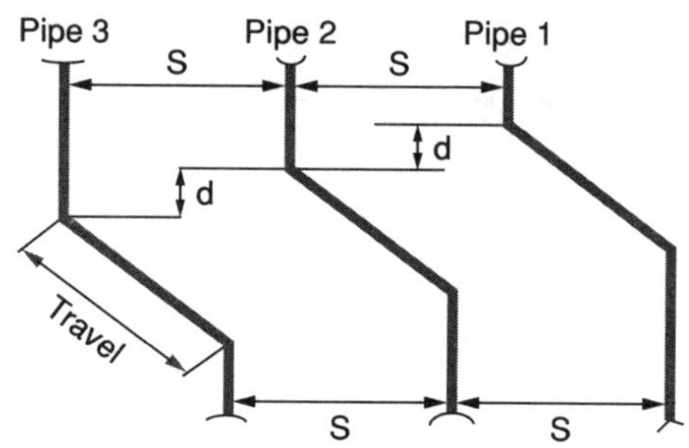

Multiple Pipe Offsets

To calculate a multiple offset using a uniform spread:

1. Calculate the spread (S).
2. Multiply S by the constant. The constant value is given in the table Multiple Pipe Offsets and is based on the fitting size.
3. Calculate the d value.
4. Measure each offset pipe d value longer than the pipe installed next to it.

Example:

Calculate two pipes 12 inches center-to-center (C–C). The offset is 26 inches with 45° elbows. Calculate the length differences and the C–C length of the travel pipes.

$$d = S \times \text{constant}$$

1. Enter known values in the equation. 12 inches is given and the constant value is located on the Multiple Pipe Offsets illustration above.

$$d = 12" \times 0.414$$

2. Calculate the *d* value. This is the difference in offset pipe lengths.

$d = 12" \times 0.414 = 4.968$
4.968 is approximately 5"

Note: Although pipe lengths change, travel remains the same for all pipes.

Rolling Offsets

Sometimes a pipe changes direction both horizontally and vertically. This is called a rolling offset. The illustration below displays a rolling offset with 45° fittings and a 60° elbow. Often a wye fitting is used instead of the 60° elbow.

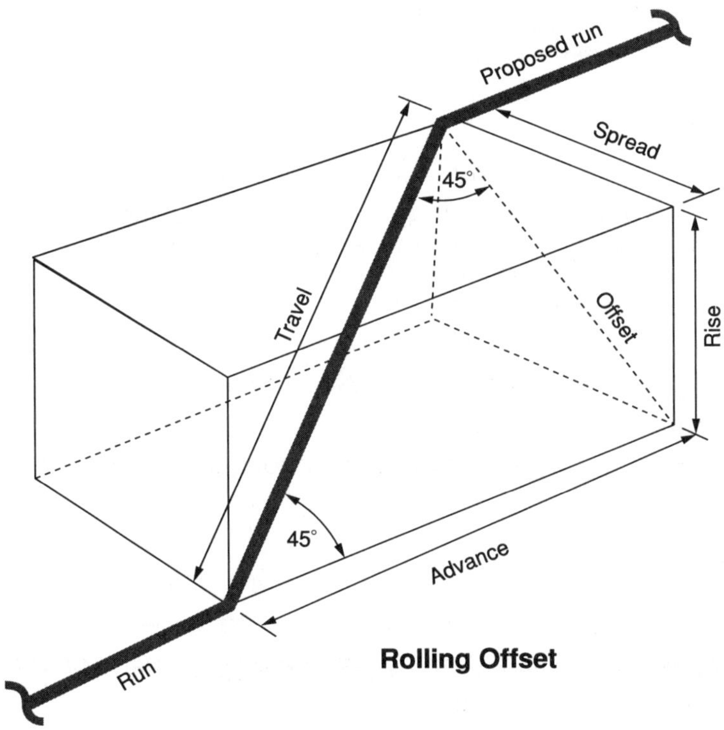

Rolling Offset

Squared values and square roots are used to calculate the offset distance. Squared values are written as 5^2 or whatever the number is $\#^2$. To calculate a squared value, take the given number and multiply the given number by itself.

$2^2 = 4$ $3^2 = 9$ $4^2 = 16$ $5^2 = 25$

The offset is calculated by using the square root of the rise and spread. Rise is the actual vertical distance the pipe is offset and the spread is the horizontal distance the pipe is offset.

$$\text{Offset} = \sqrt{\text{rise}^2 + \text{spread}^2}$$

Notes

Notes

ROLLING OFFSETS CONSTANTS		
Fitting	**Travel = Angle**	**Travel = Offset ×**
1/32	11¼°	5.126
1/16	22½°	2.613
1/8	45°	1.414
1/4	60°	1.155
1/5	72°	1.051

$$\text{Offset} = \sqrt{\text{rise}^2 + \text{spread}^2}$$

The square root of a number x is shown by $\sqrt{x}$. When multiplied by itself the square root produces the value x under the square root symbol. Use a calculator to calculate square roots quickly.

The square root of 4 is 2. $2 = \sqrt{4}$
The square root of 9 is 3. $3 = \sqrt{9}$
The square root of 25 is 5. $5 = \sqrt{25}$

A plumber must know the constant, the length of the rise, and the spread value when calculating the travel of a rolling offset.

To calculate a rolling offset:

1. Square the rise.
2. Square the spread.
3. Add the values of step 1 and step 2.
4. Calculate the square root of step 3.
5. Calculate your travel.

Example:

Calculate the travel for a 45° rolling offset with a rise of 5 inches and a spread of 13 inches.

$$\text{Offset} = \sqrt{\text{rise}^2 + \text{spread}^2}$$

1. Enter the known values in the equation, 5 inches is the given rise and the given spread is 12 inches.

$$\text{Offset} = \sqrt{5^2 + 13^2}$$

2. Calculate the squared values of the rise and spread.

$\text{Rise} = 5^2 = 25$
$\text{Spread} = 13^2 = 169$

3. Add the rise squared and the spread squared.

   $25 + 169 = 194$

4. Calculate the square root of 194.

   $13.94 = \sqrt{194}$

5. Place the known values to calculate the travel.

   Travel = Offset × 1.414 (given in Rolling Offsets Constants table)
   Travel = 13.94 (given in step 4) × 1.414

6. Calculate travel.

   Travel = 13.94 × 1.414 = 19.71
   Travel = 19.71" ≈ 19¾"

Common Rolling Offset

A common rolling offset used in plumbing is one that consists of a 45° wye and a 60° elbow. This is a very convenient set of values for a plumber. When these figures are used to implement a rolling offset, the rise and spread lengths are equal and the travel equals two times that length. For this exact situation, you can calculate the travel by multiplying the rise or the spread by 2.

Situation = Rolling offset with a 45° wye and a 60° elbow
Travel = Rise × 2 = Spread × 2

Pitch and Grade

Pitch and grade are used interchangeably. Drainage piping is set at a certain pitch (grade) to ensure adequate drainage. Pitch (P) is the slope of a line in respect to the horizontal plane. Pitch is displayed in inches per foot.

In the pipe's given length, fall (F) is quantity the line drops. Fall (drop) is expressed in inches.

Notes

Notes

Pipe Length (L) is the length of the pipe displayed in feet.

Several equations are possible when dealing with pitch.

Fall = Length × Pitch
Length = Fall ÷ Pitch
Pitch = Fall ÷ Length
Pitch = Length × % grade

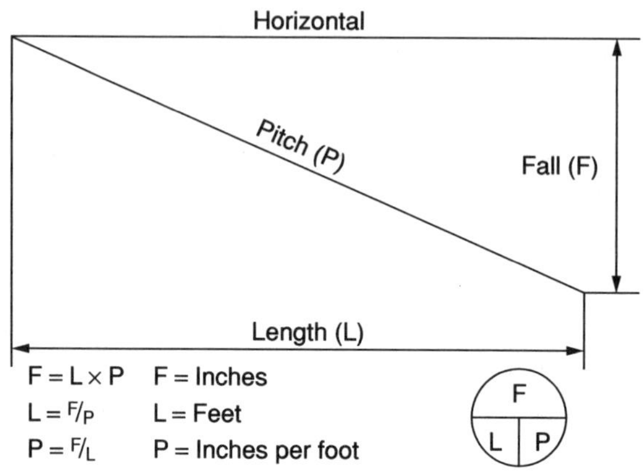

Examples:

Calculate the fall of a pipe run 32 feet long with a pitch at ¼" per ft.

Fall = Length × Pitch

1. Enter known values in the equation

   Fall = 32' × ¼"

2. Calculate the value.

   Fall = $\frac{32}{1} \times \frac{1}{4} = \frac{32}{4}$

3. Simplify the value

   Fall = $\frac{32}{4}$ = 8"

Calculate the length of a pipe with a pitch at ⅛ inch/foot and a fall of 40 inches.

Length = Fall ÷ Pitch

1. Enter the known values in the equation.

$$\text{Length} = 40" \div \tfrac{1}{8}$$

2. Calculate the value.

$$\text{Length} = 40" \times \tfrac{8}{1} = \tfrac{320}{1}$$

3. Simplify the value.

$$\text{Length} = \tfrac{320}{1} = 320'$$

Calculate the pitch of a pipe that is 160 feet long and falls at 16".

$$\textbf{Pitch} = \textbf{Fall} \div \textbf{Length}$$

1. Enter known values in the equation.

$$\text{Pitch} = 16" \div 160'$$

2. Calculate the value.

$$\tfrac{16}{160} = \tfrac{1}{10} \text{ inch/foot}$$

Converting Percentage Grade to Fractional Inches per Foot

When converting a percentage grade to the equivalent fractional grade in inches per foot, divide the percentage grade by 100. This value is multiplied by 12 to calculate the decimal figure.

$$\textbf{Pitch} = \textbf{(\% grade} \div \textbf{100)} \times \textbf{12}$$

Example:

Calculate the fractional pitch for a 3% grade.

1. Enter the known values in the equation.

$$\text{Pitch} = (3\% \div 100) \times 12$$
$$\text{Pitch} = (3 \div 100) \times 12$$

2. Calculate the value.

$$\text{Pitch} = (\tfrac{3}{100}) \times 12$$
$$\text{Pitch} = (0.03) \times 12$$

Notes

Notes

Pitch = $0.03 \times 12 = .36$
Pitch = $0.36 \approx \frac{1}{3}$ inch/foot
3% grade = $\frac{1}{3}$ inch/foot

Converting Pitch in inches per foot to Percentage Grade

When converting a fractional grade in inches per foot to the equivalent percentage grade, divide the fractional pitch by 12. This value is multiplied by 100 to calculate the percentage figure.

Percentage grade = (Pitch ÷ 12) × 100

Example:

Calculate the percentage grade for a fractional pitch of ½.

1. Enter the known value in the equation

Percentage grade = $(\frac{1}{2} \div 12) \times 100$

2. Calculate the value.

Percentage grade = $(\frac{1}{2} \div \frac{12}{1}) \times 100$
Percentage grade = $(\frac{1}{2} \times \frac{1}{12}) \times 100$
Percentage grade = $(\frac{1}{24}) \times 100$
Percentage grade = $\frac{1}{24} \times 100 = \frac{100}{24} = 0.24$

Pitch = $0.24 \approx \frac{1}{4}$ inch/foot

WATER CALCULATIONS

Water is present in three types of known states.

- Solid – Ice
- Liquid – Water
- Gas – Steam

Solid, liquid, and gas change form. At 212°F, water boils. At 32°F, water changes to ice. These degrees are calculated from sea level. Water is at its highest density at 39.1°F. When water at 39.1°F is heated or cooled, it expands. Frozen water expands to 8⅓% of its normal volume. A cubic foot of water equals 1728 cubic inches.

For example, if a cubic foot of water (1728 cubic inches) freezes, you now have 1872 cubic inches of ice. Also when water freezes, an extra 2000 pounds per square inch of pressure builds up in the pipe.

Calculating Water Pressure

Pressure is used to force water into and through buildings. Water pressure equals 0.434 pounds per square inch (psi) at the base of any container. Pressure gauges indicate the psi in the pipe to which it is connected. Most pressure gauges correct for the atmospheric pressure and are marked as pounds per square inch (psi). A head is a column of water. A one-foot head equals a container with one foot of water. A three-foot head contains three feet of water, and so on. Pipe is considered a container. For every one-foot vertical increment, 0.434 psi is exerted. Diameter is not a factor. Pressure is directly related to the length of the pipe.

2.3 feet of head equals 1 psi of water

$2.3' = 1 \div 0.434$

Given this calculation, for every foot of pipe, 0.434 psi is lost in pressure. This is a 1 psi loss for every 2.3 feet in height. To calculate the correct pressure or head for a building, you require the known head or the known pressure.

$\text{Pressure} = \text{Head} \times 0.434$

$\text{Head} = \text{Pressure} \times 2.3$

Notes

PRESSURE AND HEAD			
Head in Feet	**Pressure (psi) Pounds Per Square Inch**	**Head in Feet**	**Pressure (psi) Pounds per Square Inch**
2.304	1	1	0.434
4.608	2	2	0.868
6.912	3	3	1.302
9.216	4	4	1.736
11.520	5	5	2.170
13.824	6	6	2.604
16.128	7	7	3.038
18.432	8	8	3.472
20.736	9	9	3.906
23.040	10	10	4.340

PSI and Heads   Pressure = Head × 0.434   Head = Pressure × 2.3   Water = 62.5 pound per cubic foot

Notes

Example:

Calculate the pressure created by a head 72 feet 6 inches.

Pressure = Head × 0.434

1. Enter the known values in the equation.

Pressure = 72' 6" × 0.434
Pressure = 72.5' × 0.434

2. Calculate the value.

Pressure = 72.5' × 0.434 = 31.465
Pressure ≈ 31½ psi

Example:

Calculate the head equivalency to a pressure of 60 psi.

Head = Pressure × 2.3

1. Enter the known values in the equation.

Head = 60 psi × 2.3

2. Calculate the value.

Head = 60 psi × 2.3 = 138 feet
Head = 138 feet

Volume of a Pipe or Cylindrical Container

The volume of a pipe or cylindrical container can be calculated by multiplying 0.7854 by the diameter squared (D^2) in inches by the height (H) in feet.

Example:

Calculate the volume of pipe 14 feet long with a diameter of 4 inches.

Volume = 0.7854 × D^2 × H

Notes

1. Enter the known values in the equation.

$$\text{Volume} = 0.7854 \times 4^2 \times 14$$

2. Calculate the value.

$$\text{Volume} = 0.7854 \times 16 \times 14$$
$$\text{Volume} = 175.9296 \text{ cubic inches}$$

Volume of a Pipe Length

The volume of a given length of pipe is calculated by multiplying 0.0408 by the inside diameter squared (D^2) by the length (L) in feet.

$$\mathbf{Volume = 0.0408 \times D^2 \times L}$$

Example:

Calculate the volume for a pipe 10 feet long with a diameter of 6 inches.

1. Enter the known values in the equation.

$$\text{Volume} = 0.0408 \times 6^2 \times 10'$$

2. Calculate the value.

$$\text{Volume} = 0.0408 \times 36 \times 10'$$
$$\text{Volume} = 14.688 \text{ gallons}$$

Note: When you double the diameter of a pipe, you increase the volume four times for each unit length you are calculating.

Weight of Water in a Pipe

When calculating the weight of water in a pipe, multiply 0.34 by the diameter squared (D^2) by the length (L) in feet of the pipe.

$$\mathbf{Weight = 0.34 \times D^2 \times L}$$

Example:

Calculate the weight of water in a 6 inch pipe that is 8 feet long.

Notes

1. Enter the known values in the equation.

$$\text{Weight} = 0.34 \times 6^2 \times 8'$$

2. Calculate the value.

$$\text{Weight} = 0.34 \times 36 \times 8'$$
$$\text{Weight} = 97.92 \text{ pounds}$$

Converting Cubic Inches to Gallons of Water

231 cubic inches of water equals 1 gallon of water. To calculate gallons divide the volume of water by 231.

Volume = Cubic Inches ÷ 231

Example:

Convert the volume of a container with 5775 cubic inches.

1. Enter the known values in the equation.

$$\text{Volume} = 5775 \div 231$$

2. Calculate the value.

$$\text{Volume} = 5775 \div 231 = 25 \text{ gallons}$$

PART TWO

Math Practice Exam

Notes

DETERMINING MARKUP

If a plumber buys a water closet tank repair part for $15.00 and sells it to the customer, the plumber adds 30½%. How much will the customer be charged?

1. Enter the known values in the equation.

   $30\frac{1}{2}\% = 30.5\%$

2. Write the number as a fraction and reduce it.

   $305/1000 = 61/200$

3. Multiply the figures.

   $\$15 \times 61/200 = 915/200 = \$4.58$

4. Add the markup price.

   $15.00 Cost
   +4.58 Markup
   $19.58 Selling Price

Note: Insert the markup you desire in place of the 30½%

Percentage, Base, and Rate

You may be required to calculate a specific percentage (P), base (B) or rate (R).

$$\mathbf{P = B \times R}$$
$$\mathbf{B = P \div R}$$
$$\mathbf{R = P \div B}$$

Example:

Figure the payment percentage due on a job. Initially, a \$200 down payment was made. The contract asks for \$600 total.

1. Enter the known values in the equation.

$$P = B \times R$$
$$\$200 = \$600 \times R$$

2. Divide both sides by 600

$$200/600 = 1 \times R$$

3. Reduce the fraction

$$1/3 = R$$

4. Convert the fraction to a percentage.

$$1/3 = 0.33$$

33% was paid up-front
67% is left to be paid

Notes

Notes

PRACTICE EXAM

1. What is an Offset?

   a. The vertical distance between two pipes

   b. Pipe assemblies connecting a section of pipe from one line to another

   c. The length of pipe less the fittings

   d. All the above

2. Calculate the travel of a 45° offset with an offset of 28 inches.

   a. 39½"

   b. 33½"

   c. 11⅜"

   d. 37¼"

3. Calculate the travel of a 45° rolling offset with a rise of 8 inches and a spread of 20 inches.

   a. 30"

   b. 29¾"

   c. 30½"

   d. 33¼"

4. Calculate the fall of a drainage pipe 110 feet at a 4% grade.

   a. 4⅜"

   b. 5"

   c. 4"

   d. 4⅞"

5. Calculate the pressure created by a head 50 feet 9 inches.

   a. 20.23 psi
   b. 21 psi
   c. 24 psi
   d. 22.03 psi

6. What is the volume of a pipe 14 feet long with a diameter of 6 inches?

   a. 21.7836 gallons
   b. 20.5632 gallons
   c. 14.3268 gallons
   d. 19.8887 gallons

7. A customer wanted 40 psi at a building with only 20 psi available. The plumber was told to install a water tower. How much head would he need to have the 40 psi?

   a. 40'
   b. 17.36'
   c. 92'
   d. 86.5'

8. Fractional numbers ____________________.

   a. are part of the whole number
   b. are expressed in decimal form
   c. can be converted to decimals
   d. Both A and C

9. When installing multiple pipe offsets, what must be considered?

   a. $d = S \times$ constant
   b. $d =$ difference in offset pipe lengths
   c. Pipe lengths change but travel remains the same for all pipes
   d. All of the above

Notes

Notes

10. Whole numbers are ____________________.

    a. whole numbers and fractions
    b. positive numbers without fractions or decimals
    c. positive numbers with fractions and decimals
    d. numbers with decimals

11. To calculate a rolling offset, you must ____________________.

    a. square the rise
    b. square the spread
    c. Both A and B
    d. calculate square root of both A and B
    e. All the above
    f. None of the above

12. You are to install a new sewer line from building to sewer main. The sewer at the building is 3 feet deep. The ground surface is level. The distance from building to sewer main is 75 feet. How much deeper (fall) will the pipe need to be at the main sewer if the pitch (slope) is ⅛ inches per foot?

    a. 9⅜"
    b. 19½"
    c. 12¼"
    d. 14½"
    e. None of the above

13. The pressure at street level is 60 psi. You must run a water service to a house built on a 30 foot rise. What is going to be the pressure at the house?

    a. 60 psi
    b. 13.8 psi
    c. 34 psi
    d. None of the above

14. What is the one constant to memorize when calculating pipe offsets?

    a. 0.0408
    b. 1.414
    c. 2.3
    d. 0.434

15. When calculating pipe offsets, what must be given?

    a. Constant
    b. Offset (base)
    c. Advance (altitude)
    d. Either B or C
    e. All of the above

16. Calculate the weight of water in a 10 inch pipe that is 10 feet long.

    a. 340 lbs.
    b. 34 lbs.
    c. 340 psi
    d. 34 psi

17. Calculate volume of a tank (pipe) 20 feet in height with a diameter of 48 inches.

    a. 36,191.232 cu. in.
    b. 31,619.623 cu. in.
    c. 24,915.167 cu. in.
    d. 316,196.23 cu. in.

Notes

Notes

18. How many gallons does the tank in question 17 hold?

    a. 157.627 gal.

    b. 156.672 gal.

    c. 149.552 gal.

    d. 162.712 gal.

19. Calculate a 1½ inch pipe offset that has an advance of 48 inches.

    a. 67⅞"

    b. 72⅞"

    c. 64⅞"

    d. 67⅞"

20. You buy a pressure reducing valve (PRV) for $35.98 to install for a customer. You want to make a 33⅓% markup. What would you charge the customer?

    a. $47.96

    b. $35.98

    c. $48.00

    d. $62.36

21. Why are labels on pipe, valves, etc. important?

    a. Labels identify hazards

    b. Labels identify contents

    c. Labels identify units of measurements

    d. All of the above

22. Your estimate totals $4,450.00 to replace a 4 inch sewer line. You want 60% down and 40% balance upon completion of job. How much do you need to collect before you start the job?

  a. $2,225.00

  b. $2,670.00

  c. $1,780.00

  d. None of the above

23. You estimated a job for $2,195.00. You agree to $900.00 up front with the balance paid when job is completed. What is the percentage you will collect up front?

  a. 33⅓%

  b. 45%

  c. 41%

  d. 38½%

24. What is the balance due at the completion of the job from Question 23?

  a. $1,295.00

  b. $900.00

  c. Neither A nor B

  d. Both A and B

25. What does a *fraction bar* mean?

  a. Multiply

  b. Add

  c. Subtract

  d. Divide

Notes

Notes

PART THREE
Pipe, Fittings and Symbols

This section serves as a review of pipe, fittings and symbols used throughout most plumbing exams. A review of this section will prove beneficial in understanding the code, math, general knowledge, and identification practice exams that follow.

Notes

PIPES

Drainage Pipe

Drainage pipes encountered by plumbers might be cast iron, steel, galvanized, DWV copper, schedule 40 PVC DWV plastic, clay, cement, or lead. The most common pipe used today is schedule 40 PVC because it is easy to work with and it is economical.

Cast Iron: Comes in single hub 5' and 10' and 5' double Hubbed. Also comes in standard and extra-heavy weights. Color is black.

Steel: Usually comes in 20- or 21-foot lengths. Color is black.

Galvanized Steel: Usually comes in 21-foot lengths. Color is black.

DWV Copper: Usually comes in 20-foot lengths. Very expensive. Color is copper with a yellow stripe lengthwise.

PVC Schedule 40 DWV: Comes in various lengths. Color is white.

Plastic: A lightweight pipe used for leach field lines. Comes with or without holes for drainage. Comes in different colors, mostly white.

Clay: Used mostly in main sewer and storm drainage systems

Cement—used mostly in storm drainage systems

Note: Lead pipe is now prohibited because of the potential hazard of lead poisoning.

Water Distribution Pipe

Water distribution pipes that you will encounter might be galvanized steel, Copper types K, L, and M, schedule 40 PVC, CPVC, PE (polyethylene), PB (polybutylene), and PEX.

Copper Type K: Heaviest gauge used in commercial, industrial and for underground water service lines. Color code is green.

Copper Type L: Most used of the copper types. Can be used for hot or cold water lines. Comes in various sizes ¼" to 4", also various lengths, 2' to 20'. Color code is blue.

Copper Type M: Used for inside building and in walls. Not to be used in the ground, very thin. Comes in various sizes ½" to 2", and various lengths 5' to 20'. Color code is red.

PVC Schedule 40: Used for cold water distribution only. Can be used for water service lines. Comes in various sizes ½" to 4", and comes in various lengths, 5' to 20'. Color is white.

CPVC: Used for hot and cold water distribution. Comes in various sizes ½" to 2", and various lengths 5' to 20'. Color is cream.

PE (Polyethylene): Usually sold in coils. Used for water supply systems, jet wells, farm sprinkler systems, chemical well lines, and gas gathering systems. Joined together with plastic insert fittings and clamps. Sold in sizes ¾", 1", 1½", and 2". Color is black.

PB (Polybutylene): Flexible tubing (old style was gray—new style is milky white). Comes in 20-foot lengths and 50 to 1000 foot coils. Used for residential water supply only, mostly in mobile homes.

PEX: One of the most versatile, easy-to-use plumbing systems available. Can be used for hot- or cold-water systems. Uses brass fittings and copper crimp rings. A manifold can be used which has a shut-off valve for each line. Color usually is white. Can be color red for hot and color blue for cold.

Note: Be sure and check local codes. Some types of pipe may be prohibited in your area.

Notes

Notes

FITTINGS

Elbows: Should be read as follows:

$\frac{1}{16}$ bend or $22\frac{1}{2}°$ bend
$\frac{1}{8}$ bend or 45° bend
$\frac{1}{4}$ bend or 90° bend
$\frac{1}{6}$ bend or 60° bend
$\frac{1}{5}$ bend or 72° bend

Tees and Wyes: Should be read by reading the flow first then the side inlet. If it is a reducing fitting, the largest flow side should be read first.

Example:

A tee that has a flow of ¾" and a side inlet of ½" would be read: ¾ × ¾ × ½ tee, but, when the flow sizes are the same, you read it: ¾ × ½ tee. The second ¾ is understood.

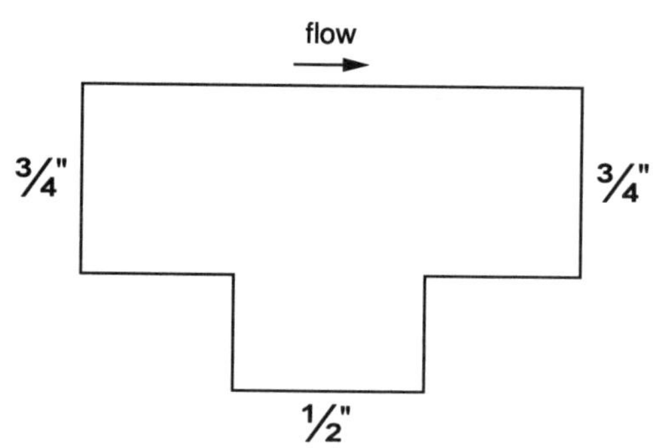

Example:

The same rules apply to reading drainage fittings:

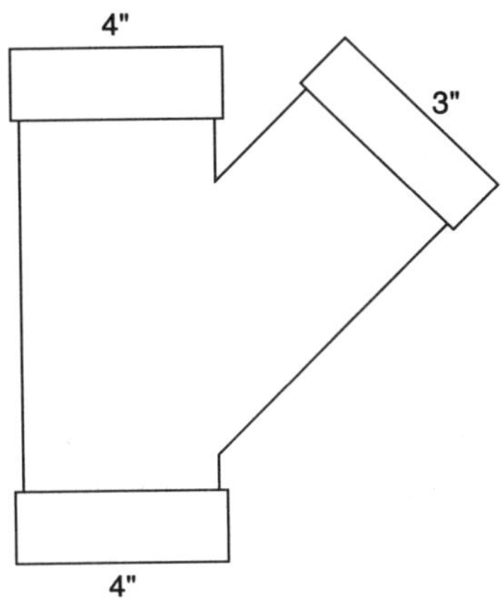

Read: 4 × 3 reducing wye

Drainage Fittings

Cast-iron fittings

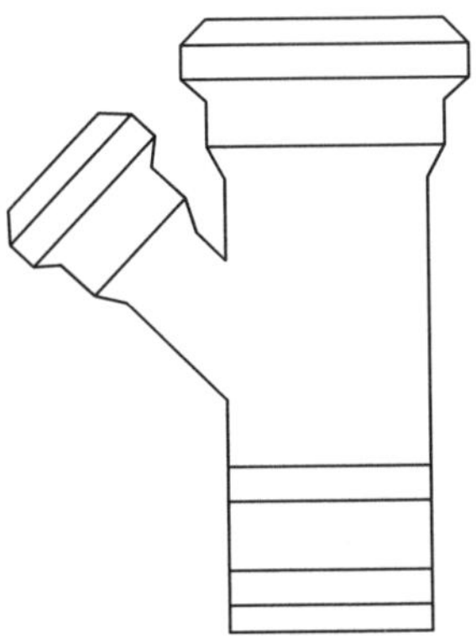

Reducing wye

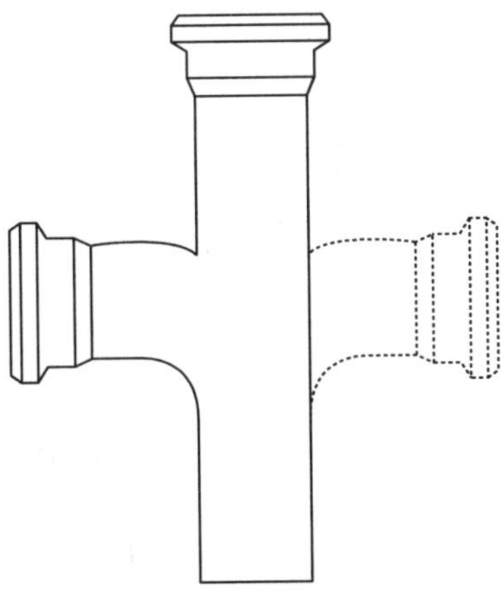

Sanitary tee

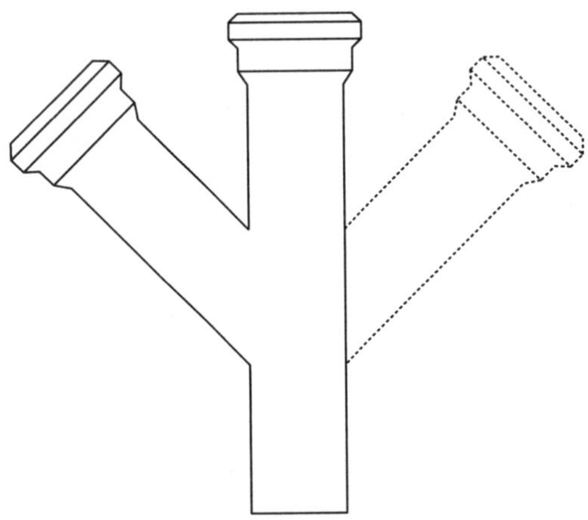

Wye

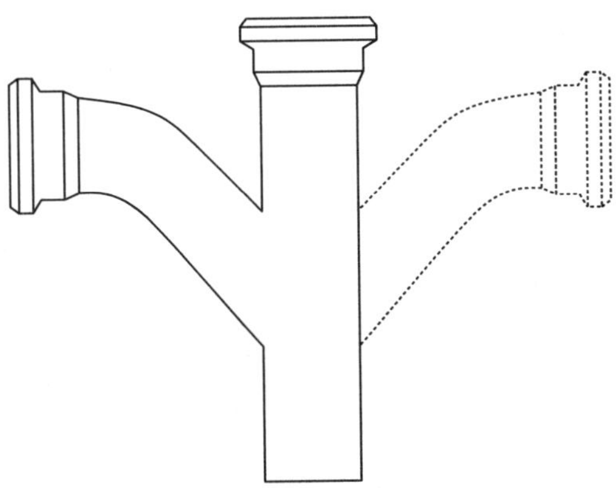

Tee wye

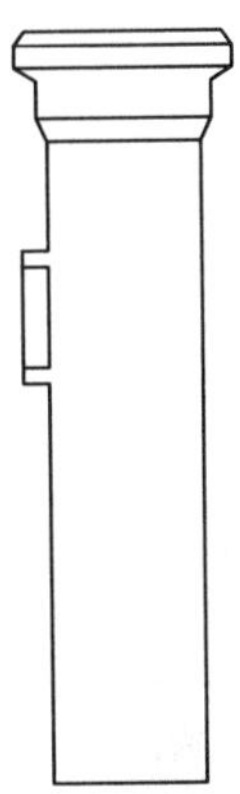

Tapped tee

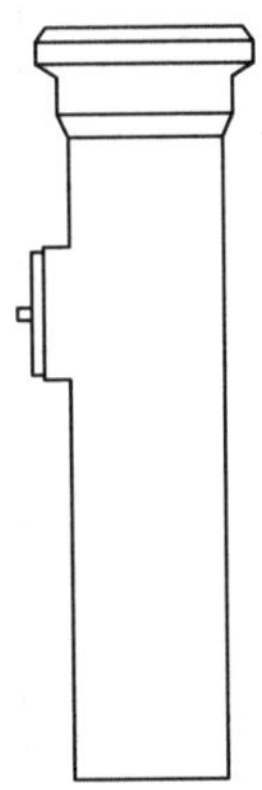

Tee cleanout

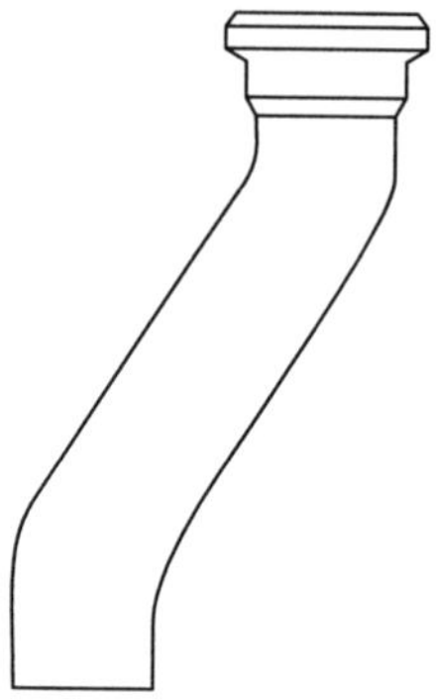
Offset

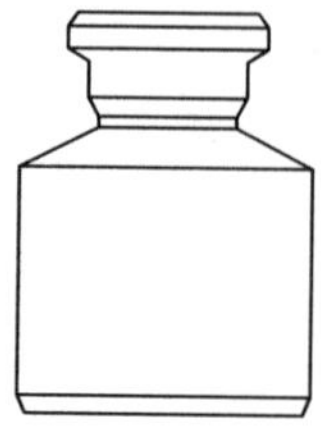
Reducer

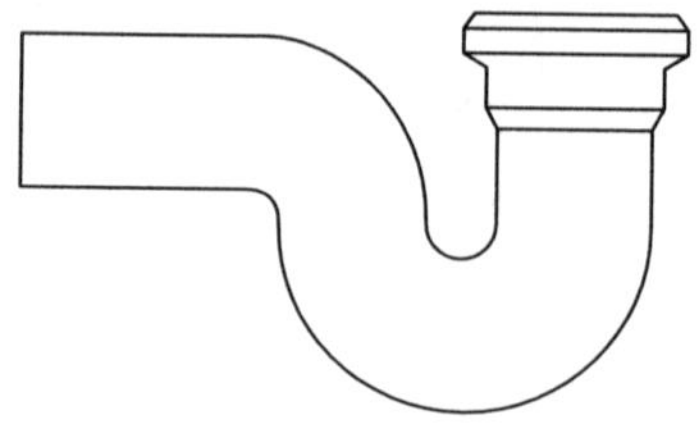
P-trap

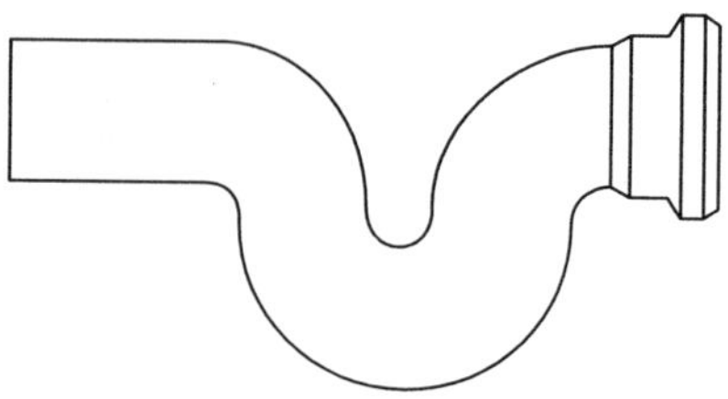
Running trap

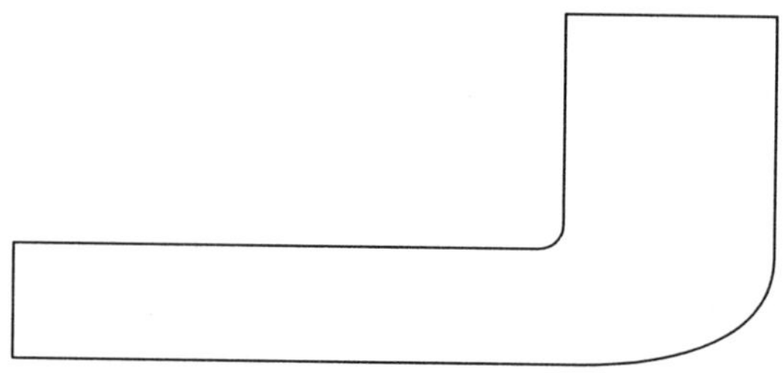
Closet bend

PVC Schedule 40 DWV Fittings

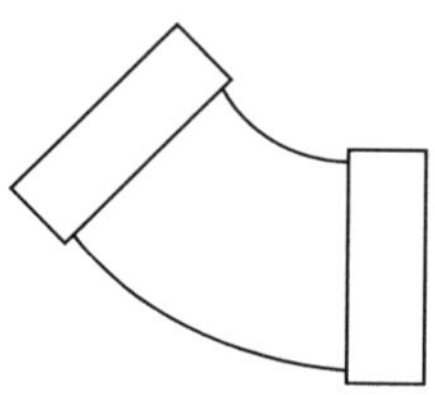
45° elbow

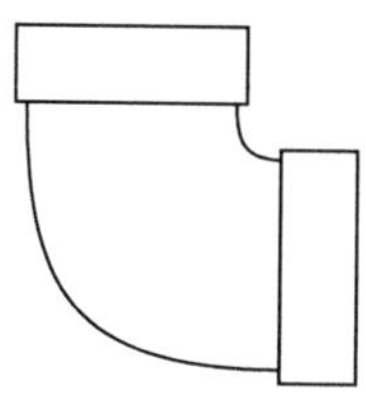
90° elbow

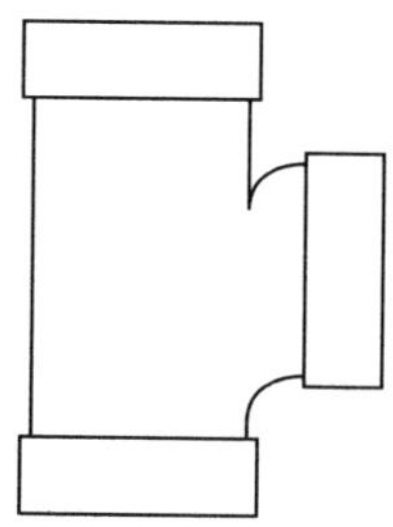
Tee

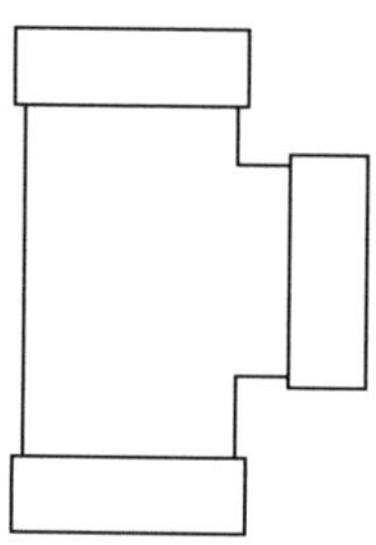
Vent tee

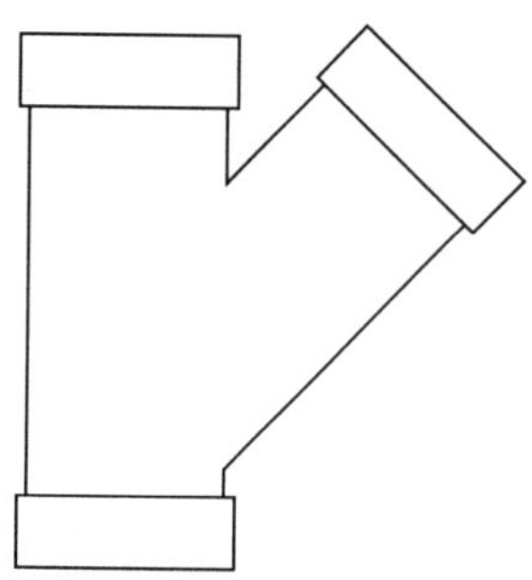
Wye

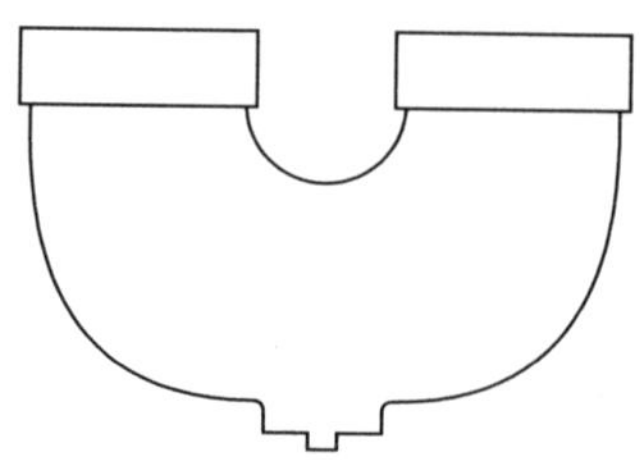
Trap

WATER DISTRIBUTION FITTINGS

Galvanized Fittings

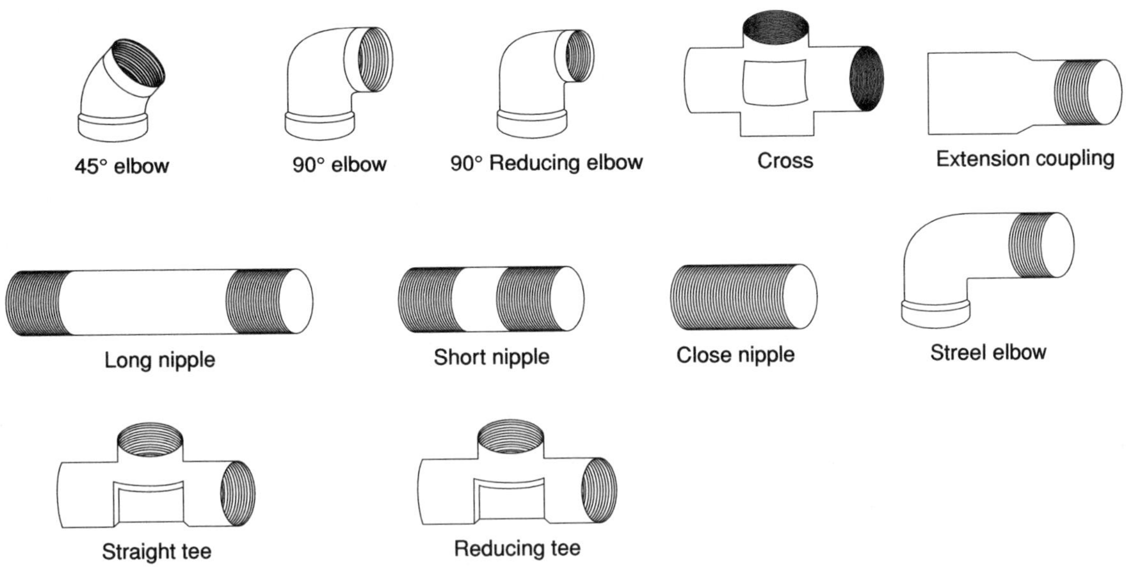

Copper Fittings

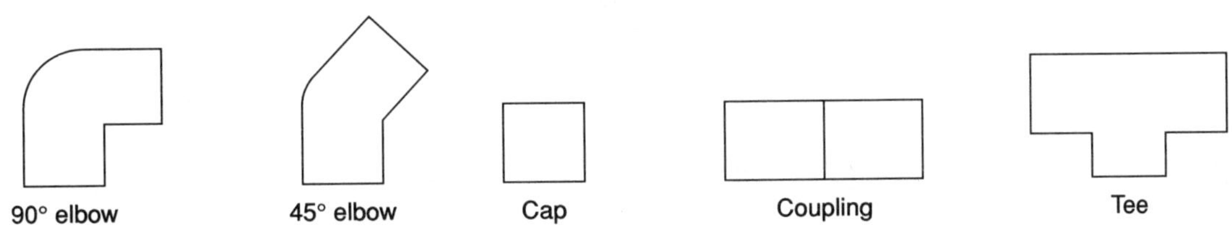

CPVC Fittings

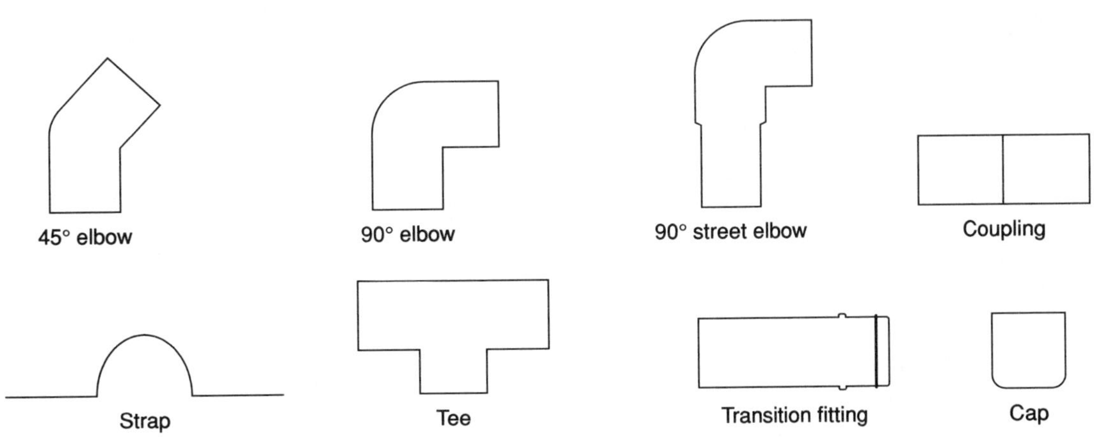

PE (Polyethylene) Fittings

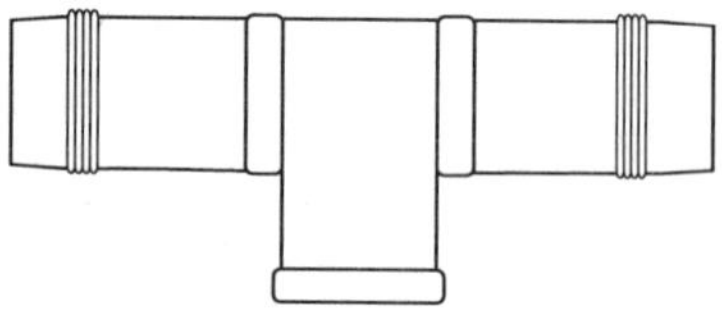
Adapter tee

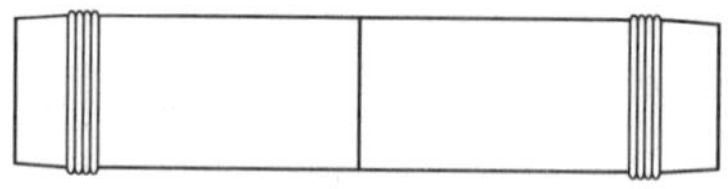
Coupling

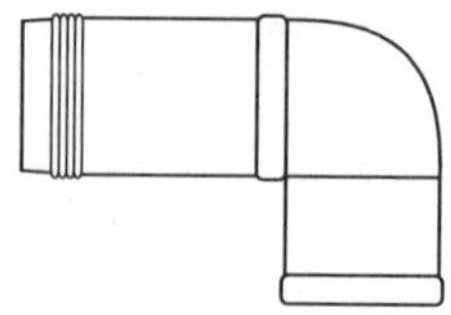
Female adapter

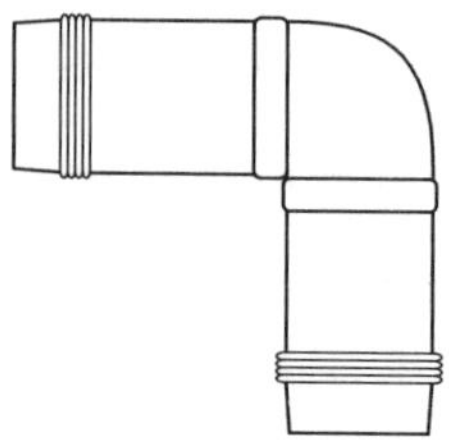
Elbow

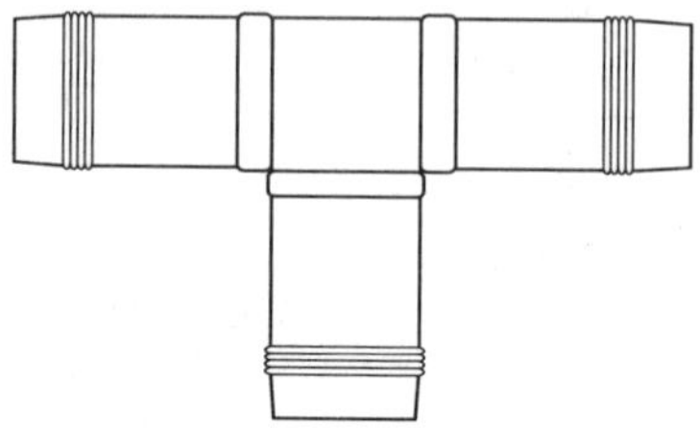
Tee

PB (Polybutylene) Fittings

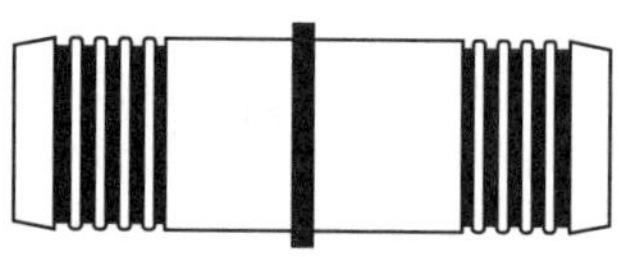
Coupling

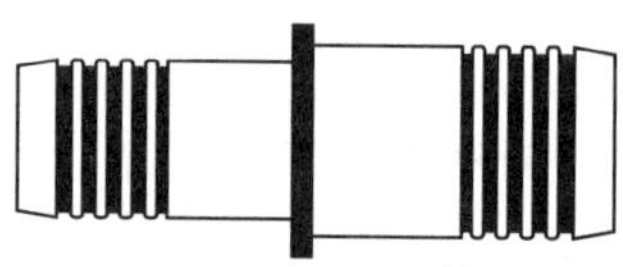
Reducing coupling

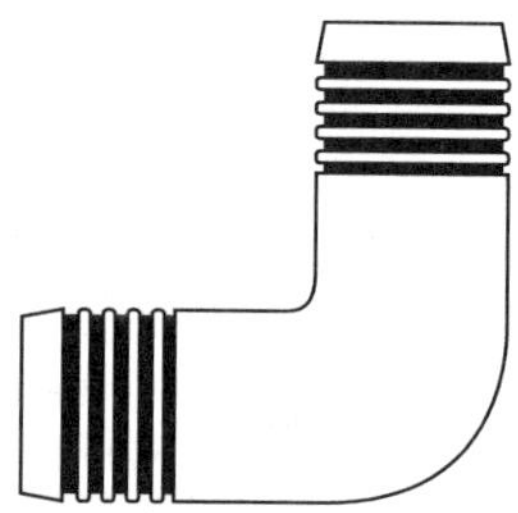
Elbow

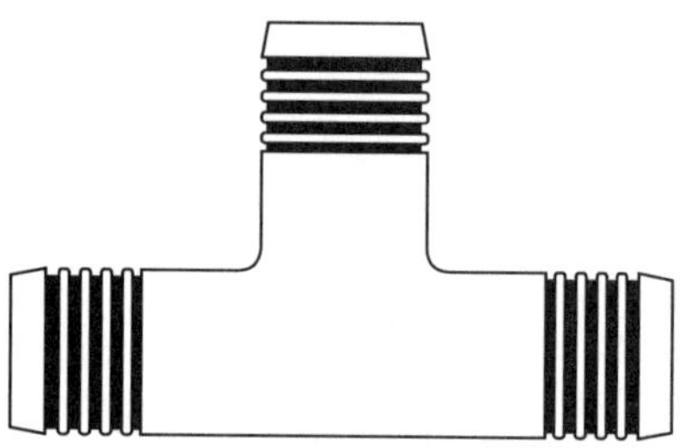
Tee

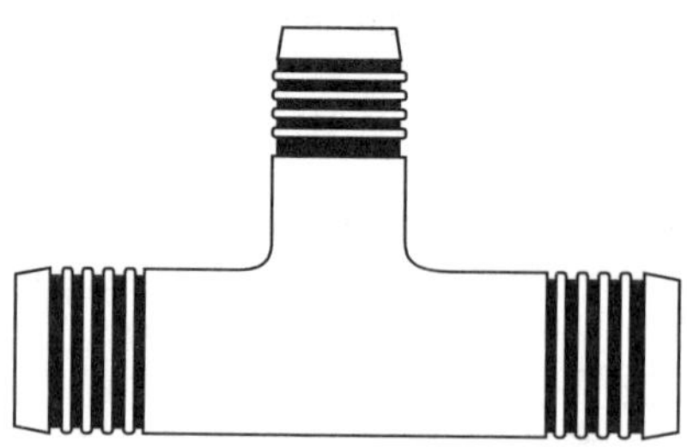
Reducing tee

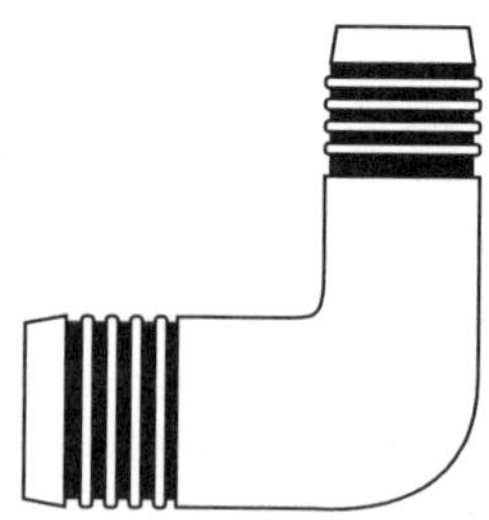
Reducing elbow

PLUMBING/BLUEPRINT SYMBOLS

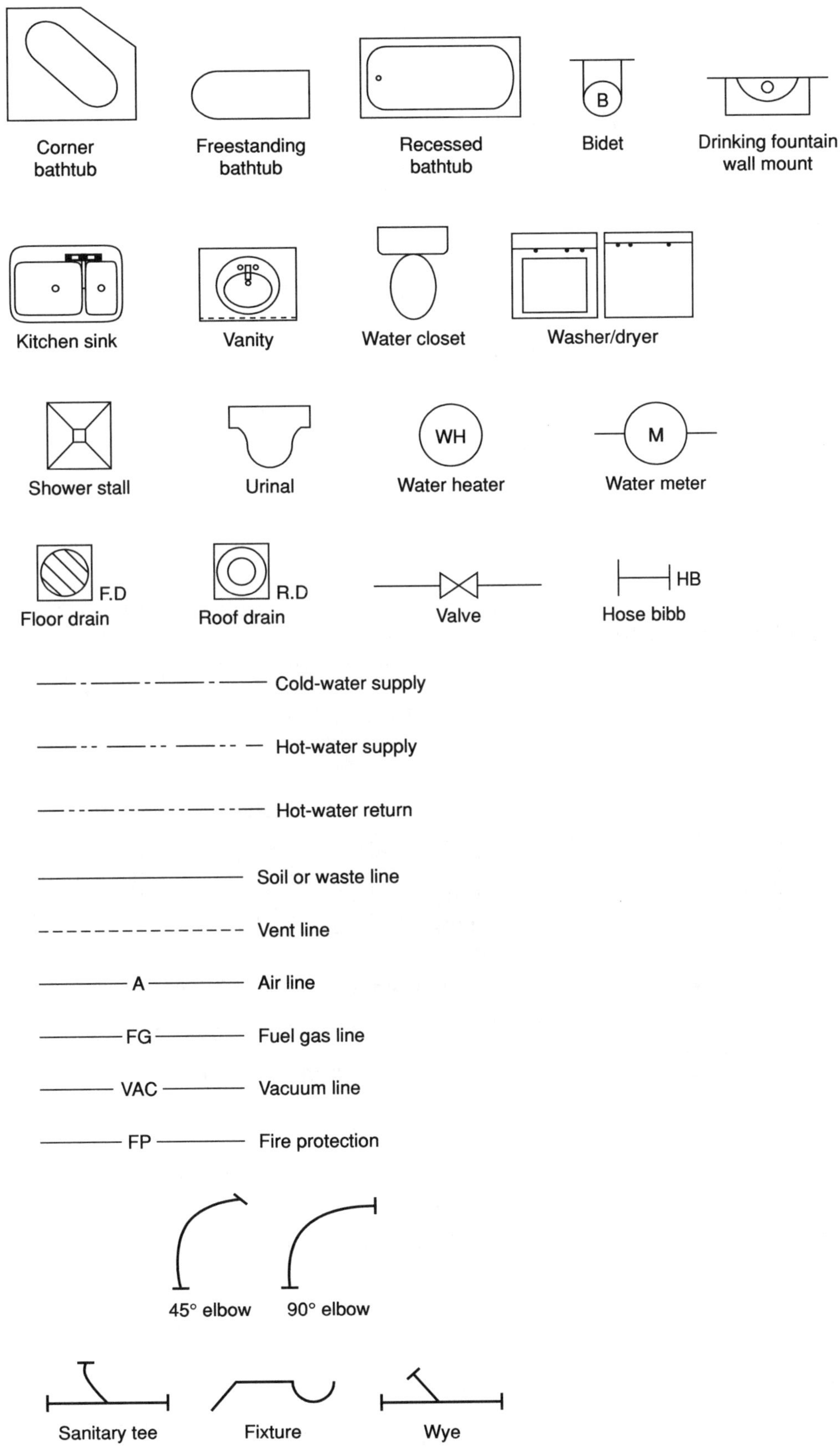

Notes

PIPE, FITTINGS AND SYMBOLS QUESTIONNAIRE

1. Which elbow(s) are read correctly?

   a. $\frac{1}{4}$ bend or 90° bend
   b. $\frac{1}{8}$ bend or 45° bend
   c. $\frac{1}{16}$ bend or $22\frac{1}{2}$° bend
   d. All of the above

2. Identify this blueprint symbol.

   a. Vanity
   b. Water closet
   c. Bidet
   d. Toilet

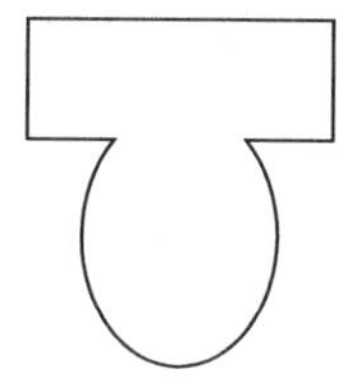

3. Where can you use PVC Schedule 40 water pipe?

   a. All water distribution systems
   b. Cold-water distribution systems
   c. Hot-water distribution systems
   d. Hot-water heating systems

4. Where can you use CPVC water pipe and fittings?

   a. All water distribution systems
   b. Cold-water distribution systems
   c. Hot-water distribution systems
   d. All of the above

5. Identify this blueprint symbol.

   a. Shower stall
   b. Washer
   c. Dryer
   d. Laundry sink

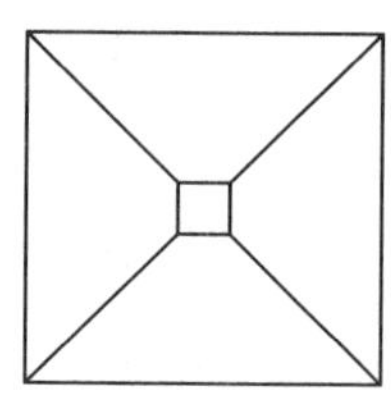

6. Identify this plumbing symbol. 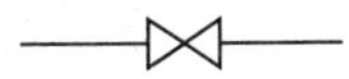

    a. Tee

    b. Wye

    c. Valve

    d. Cross

7. Some of the pipes used for water distribution are?

    a. CPVC

    b. Copper

    c. PEX

    d. PVC DWV

    e. A, B, and C

8. Drainage pipe/s you might encounter is/are.

    a. CPVC

    b. PEX

    c. PE

    d. None of the above

9. What does this blueprint symbol represent?

    a. Air-line

    b. Cold-water line

    c. Hot-water line

    d. Fire-protection line

10. Identify this fitting.

    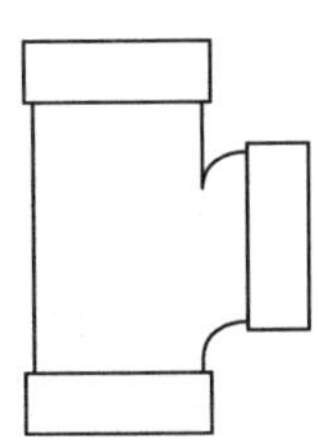

    a. Cleanout tee

    b. Sanitary tee

    c. Vent tee

    d. None of the above

Notes

Notes

11. This fitting is used with what kind of pipe?

    a. PB (polybutylene)
    b. CPVC
    c. PVC
    d. Copper

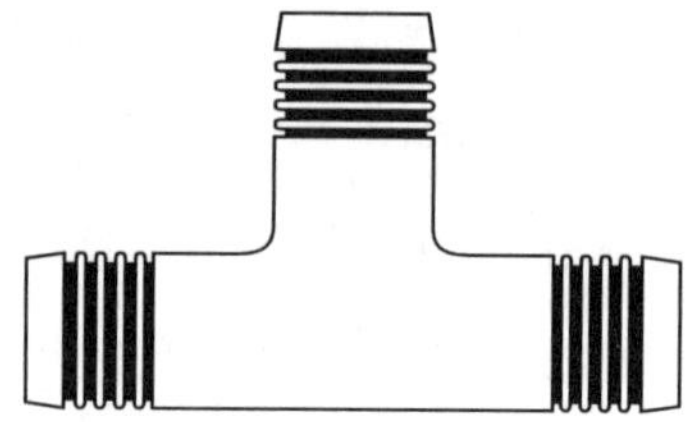

12. What does this blueprint symbol represent?

    a. Cold-water line
    b. Hot-water return line
    c. Hot-water line
    d. Air-line

13. What does this blueprint symbol represent? —— FG —— FG ——

    a. Fuel gas line
    b. Fuel gathering line
    c. Fire prevention line
    d. Drainage pipe line

14. Identify this fitting.

    a. Long turn elbow
    b. 90° bend
    c. Closet bend
    d. ¼ bend

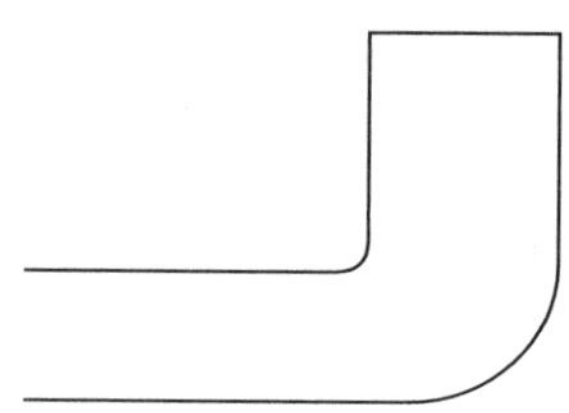

15. Identify this fitting.

    a. P-trap
    b. Running trap
    c. S-trap
    d. Drum trap

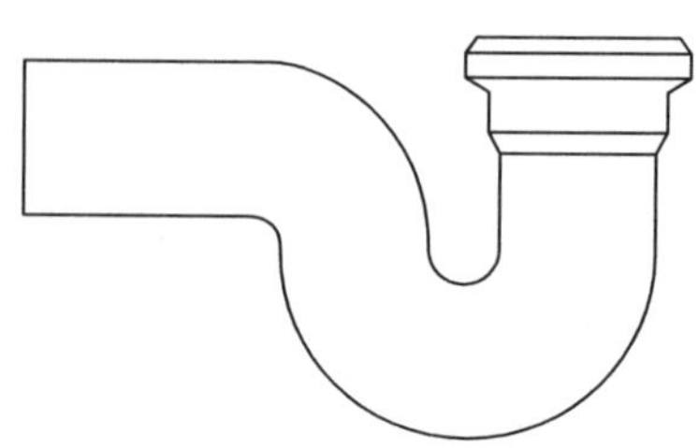

16. What are the four types of copper?

 a. Water, drainage, gas, vent

 b. K, L, M and DWV

 c. Gauge 8, 10, 12 and 14

 d. None of the above

17. How do you read the this tee?

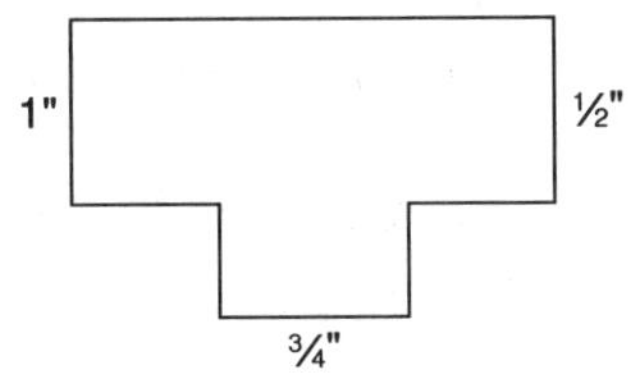

 a. 1 × ¾ tee

 b. 1 × ¾ × ½ tee

 c. 1 × ½ × ¾ tee

 d. ¾ × ½ × 1 tee

18. Which water distribution pipe is one of the most versatile, uses brass fittings and copper crimp rings, and is the easiest to use for plumbing systems today?

 a. CPVC

 b. PE

 c. PVC

 d. PEX

19. How do you read this fitting?

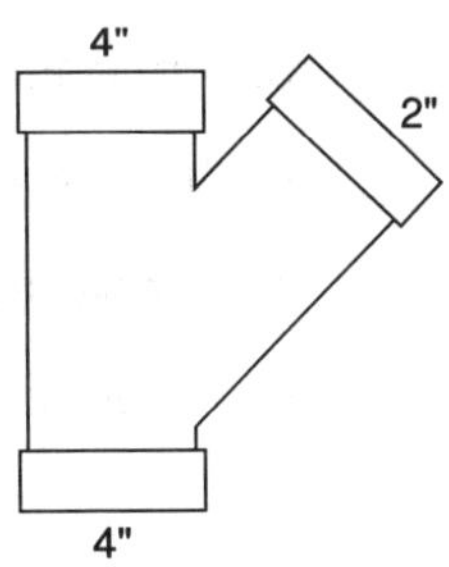

 a. 4" × 2" sanitary tee

 b. 4" × 2" wye

 c. 4" × 4" × 2 wye

 d. Either B or C

20. What can PE (polyethylene) pipe be used for?

 a. Water supply systems

 b. Jet wells

 c. Both A and B

 d. Drainage systems

Notes

Notes

21. What is this fitting called?

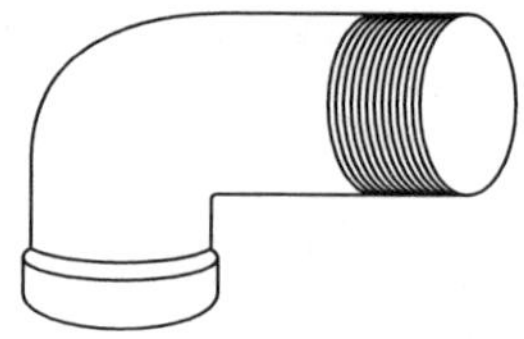

a. Street ell
b. 90° elbow
c. Both A and B
d. 45° elbow

22. What is this fitting called?

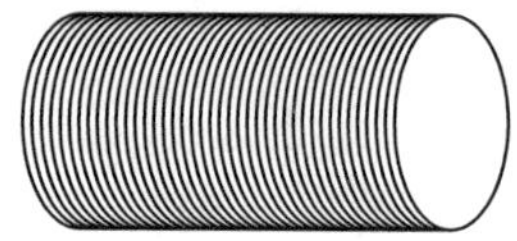

a. Short nipple
b. Close nipple
c. Short pipe
d. All of the above

23. What is most common used Drainage pipe today?

a. Copper DWV
b. Clay DWV
c. Cement DWV
d. PVC Schedule 40 DWV

24. Type M copper water pipe can be used ______________.

a. inside the building
b. underground
c. Both A and B
d. None of the above

25. When installing water systems you should ______________?

a. check with local codes
b. install the cheapest material available
c. install good quality material
d. Both A and C

PART FOUR
International Plumbing Code Practice Exams

The following section includes six practice exams based on the International Plumbing Code. You will also find several work sheets in this section outlining the steps involved to properly solve various equations necessary when using Chapter 7 of the *International Plumbing Code*. By completing each of the practice exams, you will become very familiar with the way the code book is formatted. Chapters 1 and 2 of the *International Plumbing Code* are not covered due to the fact that they cover administrative issues and definitions. However, it is recommended that you look through each of these chapters.

Notes

IPC PRACTICE EXAM #1

The following questions are based on Chapters 3 and 4 of the International Plumbing Code book and covers general regulations, fixtures, faucets, and fixture fittings.

1. What is the minimum wall thickness of the protective sheathing, which protects pipes passing through concrete walls against corrosion?

   a. ⅛"
   b. ¼"
   c. ⅜"
   d. ½"

2. Exterior water supply system piping shall be installed not less than ______ inches below the frost line.

   a. 6
   b. 12
   c. 18
   d. 24

3. Sheet lead for pans shall not weigh less than ______ pounds per square foot.

   a. 1
   b. 2
   c. 3
   d. 4

4. CPVC pipe 2 inches in diameter must be horizontally supported every ______ feet.

   a. 3
   b. 4
   c. 10
   d. 15

5. What is the maximum number of urinals allowed to substitute ten water closets in a men's restroom?

   a. 5

   b. 6

   c. 7

   d. 8

6. What is the minimum size drain required for a floor drain?

   a. 1½"

   b. 2"

   c. 2½"

   d. 3"

7. Which of the following products or materials is required to be third party tested?

   a. Plumbing appliances

   b. Backflow prevention devices

   c. Plumbing fixtures

   d. Water distribution safety devices

8. A 10,000 sq. ft. museum has a maximum occupant load of 1,500 people. What is the minimum number of water closets required for female patrons?

   a. 8

   b. 10

   c. 12

   d. 23

9. Trenches installed parallel to footings shall not extend below the ______ bearing plane of the wall.

   a. 20 degree

   b. 45 degree

   c. 60 degree

   d. 90 degree

Notes

Notes

10. A dishwasher shall be protected against backflow by a(n) ______.

    a. air break
    b. air gap
    c. vacuum breaker
    d. check valve

11. Which of the following can be used to waterproof joints around vent pipes?

    a. Lead
    b. Galvanized steel
    c. Aluminum
    d. All of the above

12. 1½ inch copper tube shall have a maximum horizontal hanger spacing of ______ feet.

    a. 6
    b. 10
    c. 12
    d. 15

13. Patrons shall be provided with public toilet facilities and have a path of travel not to exceed a distance of ______.

    a. 300'
    b. 400'
    c. 500'
    d. 600'

14. Shower compartments shall have a minimum cross-sectional area of ______.

    a. 900 sq. ft.
    b. 900 sq. in.
    c. 70 sq. in.
    d. 30 sq. in.

15. Coin-operated laundries shall be provided with floor drains having a minimum outlet of ______ inches.

    a. 2
    b. 3
    c. 4
    d. None of the above

16. What is the minimum slope required for the discharge of liquid combustion by products from condensing appliances?

    a. 1%
    b. 2%
    c. 3%
    d. 8%

17. Which of the following pipes is not required to be protected by shield plates?

    a. Copper pipe
    b. PVC
    c. Galvanized steel
    d. Brass pipe

18. What is the minimum number of plumbing fixtures required in a bar with a maximum occupancy load of 500 people?

    a. 22
    b. 23
    c. 24
    d. 25

Notes

Notes

19. 15 water fountains are required to be installed in a facility. What is the maximum number of bottled water dispensers allowed to substitute these fountains?

    a. 6

    b. 7

    c. 8

    d. 9

20. Trough urinals are ____________.

    a. required

    b. permitted

    c. allowed only when approved by an official

    d. prohibited

21. Which of the following is used to support piping?

    a. Hangers

    b. Anchors

    c. Supports

    d. All of the above

22. Shower receptors shall be constructed of __________ materials.

    a. Impervious

    b. Non-absorbent

    c. Noncorrosive

    d. All of the above

23. A 2" pipe passing through a foundation wall requires protection from a sleeve of at least ______ inches.

    a. 2

    b. 4

    c. 6

    d. 8

24. Marble vanity tops with an integral lavatory shall conform to ______________.

a. CSA B45.1

b. CSA B45.2

c. CSA B45.4

d. CSA B45.5

25. In a concealed location, piping shall be protected by a steel plate and shall extend a minimum of ______ inches above sole plate and below top plates.

a. ½

b. ¾

c. 1

d. 2

Notes

Notes

IPC PRACTICE EXAM #2

The following questions are based on Chapters 5 and 6 from the International Plumbing Code book and cover water heaters, water supply and distribution.

1. Disregarding exceptions, all of the following are considered individual water supplies except ___________.

   a. drilled wells
   b. dug wells
   c. streams
   d. land cisterns

2. What is the maximum distance a pan drain can terminate above the adjacent ground surface?

   a. 6"
   b. 12"
   c. 18"
   d. 24"

3. What is the minimum flow pressure of a temperature-controlled shower?

   a. 8 psi
   b. 15 psi
   c. 20 psi
   d. 25 psi

4. Water service pipe and the building sewer are required to be separated by a minimum of _________ feet.

   a. 5
   b. 8
   c. 10
   d. 12

5. The overflow for a water supply tank shall be covered by a corrosion-resisted screen of not less than ________ mesh per inch.

   a. 12 × 20
   b. 16 × 20
   c. 18 × 20
   d. 20 × 20

6. Hospital fixtures shall be protected against backflow with all of the following except?

   a. Reduced pressure principle backflow preventer
   b. Dual-check valve
   c. Spill-proof vacuum breaker
   d. Air gap

7. When tankless water heaters are intended for domestic use, the temperature of the water shall not exceed ________ degrees.

   a. 140
   b. 160
   c. 180
   d. 200

8. Water heaters operating above the atmospheric pressure shall be provided with ________.

   a. A levered pressure relief valve
   b. A temperature valve
   c. A self closing pressure relief valve
   d. A levered pressure relief valve and temperature valve

Notes

Notes

9. Pumps installed in basements shall be mounted on block not less than __________ inches above the floor.

   a. 6

   b. 12

   c. 18

   d. 24

10. Joints between different materials shall be made with which of the following:

    a. Dielectric fitting

    b. Approved adapter fitting

    c. Mechanical joint

    d. All of the above

11. Valves used in a water supply system shall have a maximum of __________ lead content.

    a. 0.2%

    b. 2%

    c. 4%

    d. 8%

12. Which of the following backflow prevention devices may be used in a high hazard application?

    a. Double check backflow prevention assembly

    b. Backflow preventer with intermediate atmospheric vent

    c. Barametric loop

    d. Dual-check valve

13. Hazardous piping systems must be identified and repeated at intervals not to exceed __________ feet.

    a. 25

    b. 30

    c. 50

    d. 75

14. Attics containing water heaters shall be provided with a passageway not less than ____________.

    a. 30" × 30"
    b. 22" × 30"
    c. 20" × 30"
    d. 20" × 22"

15. Which of the following fixtures require a minimum ½" supply pipe?

    a. Bidet
    b. Drinking fountain
    c. Flushing rim sink
    d. Wall hydrant

16. What is the maximum heat loss allowed by an insulated, unfired hot water storage tank is ___________.

    a. 15 Btu/h
    b. 25 Btu/h
    c. 30 Btu/h
    d. 40 Btu/h

17. What is the minimum water service pipe?

    a. ½"
    b. ¾"
    c. 1"
    d. 2"

18. Which of the following joints is not permitted between polyethylene plastic pipe and tubing?

    a. Flared joints
    b. Threaded joints
    c. Heat fusion joints
    d. Mechanical joints

Notes

Notes

19. Shut-off valves are required to be installed in all of the following locations except for ______________________.

    a. kitchen sinks in a single family dwelling

    b. bathtubs in single family dwellings

    c. water supply pipe to each sillcock

    d. water supply pipe to each appliance

20. What is the maximum flow pressure for a private lavatory?

    a. 2 psi

    b. 8 psi

    c. 15 psi

    d. 60 psi

21. Which of the following water service pipes must comform to ASTM D 2672?

    a. Polybutylene plastic pipe

    b. Ductile iron water pipe

    c. Polyvinyl chloride plastic pipe

    d. Chlorinated polyvinyl chloride plastic pipe

22. Joints between different materials shall be made with which of the following?

    a. Dielectric fitting

    b. Approved adapter fitting

    c. Mechanical joint

    d. All of the above

23. What size overflow pipe is required for a water supply tank if the supply line to the tank has a capacity of 150 to 200 gpm?

a. 2"

b. 2½"

c. 3"

d. 4"

24. Cross connections between a private water supply and public supply shall be ____________.

a. permitted

b. required

c. prohibited

d. None of the above

25. A 3 inch drain pipe can empty a water tank with a capacity of ____________ gallons.

a. 6,000

b. 4,000

c. 2,000

d. 1,000

Notes

Notes

IPC PRACTICE EXAM #3

The following questions are based on Chapter 7 of the International Plumbing Code book and covers sanitary drainage.

1. What is the maximum temperature wastewater can be when discharged into the building drainage system?

   a. 60°F

   b. 140°F

   c. 180°F

   d. 200°F

2. Brass cleanout plugs shall be used with ______________.

   a. metallic drains

   b. waste piping

   c. vent piping

   d. All of the above

3. A vertical pipe to which connections are made from the fixture side of a trap and through which vapor is removed from a device utilized on bed pan washers is known as ______________.

   a. stack vent

   b. vent stack

   c. individual vent

   d. local vent stack

4. A horizontal stack offset is located 5 branch intervals below the top of the stack. A horizontal branch interval shall not connect within ______ feet above such offset.

   a. 1

   b. 2

   c. 3

   d. 4

5. What is the minimum size sterilizer vent stack serving an instrument washer?

   a. 1"

   b. 1½"

   c. 2"

   d. 2½"

6. What size pipe is required for a building drain on 1⁄16" slope accepting discharge from 2,700 fixture units?

   a. 6"

   b. 8"

   c. 10"

   d. 12"

7. Joints between a cast iron and brass pipe shall be made with all of the following except ____________________.

   a. welded joint

   b. threaded joint

   c. caulked joint

   d. approved adapter fitting

8. A quarter bend is used as a change of direction from vertical to horizontal. What is the maximum size fitting allowed?

   a. 1½"

   b. 2"

   c. 3"

   d. 4"

9. What is the minimum slope for a 5 inch horizontal drain pipe?

   a. ½"

   b. ¼"

   c. ⅛"

   d. 1⁄16"

Notes

Notes

10. Which of the following types of pipe shall not be used for underground building drainage?

    a. Cast iron pipe
    b. Brass pipe
    c. DWV
    d. ABS

11. Soldering bushings shall be of ______________.

    a. red brass
    b. stainless steel
    c. plastic
    d. borosilicate glass

12. How many cleanouts are required for each 40 foot section of drainage pipe?

    a. 1
    b. 2
    c. 3
    d. 0

13. What is required to be installed on the discharge piping from the sewage pump in a single family dwelling?

    a. Access pit
    b. Full open valve
    c. Check valve
    d. None of the above

14. A 4 inch horizontal drain can accommodate a maximum of _________ fixture units.

a. 160

b. 170

c. 360

d. 400

15. What is the minimum capacity of a sewage ejector with a 2½" discharge pipe?

a. 21 gpm

b. 30 gpm

c. 46 gpm

d. 50 gpm

16. A sump pit shall have a diameter at least ________ inches.

a. 18

b. 24

c. 36

d. 48

17. Solvent cement joints shall be permitted ________________.

a. above ground

b. below ground

c. Both above and below ground

d. None of the above

Notes

Notes

18. What is the drainage fixture unit value for a bidet?

    a. ½

    b. 1

    c. 2

    d. 3

19. Building sewers with manholes shall be located not more than ________ feet apart from the junction of the building sewer.

    a. 100

    b. 200

    c. 300

    d. 400

20. What is the fixture unit value for an unlisted fixture with a 3" trap?

    a. 2

    b. 3

    c. 4

    d. 5

21. What is the minimum size cleanout required for a 3" drain pipe?

    a. 1"

    b. 2"

    c. 3"

    d. 4"

22. A fixture having a continuous flow of 3 gpm is equivalent to how many fixture units?

   a. 3

   b. 4

   c. 5

   d. 6

23. What is the minimum trap size for a water fountain?

   a. 1¼"

   b. 1½"

   c. 2"

   d. 2½"

24. What is the minimum clearance for rodding a cleanout of a 4 inch pipe?

   a. 18"

   b. 24"

   c. 30"

   d. 36"

25. What is the total fixture unit value for 6 floor drains and 3 public water closets rated 1.6 gpf?

   a. 21 dfu

   b. 24 dfu

   c. 30 dfu

   d. 36 dfu

Notes

Notes

Drainage Fixture Units

The following information is based on Chapter 7 of the International Plumbing Code book and covers calculations used for computing drainage fixture unit values, horizontal drains and building drains.

Drainage Fixture: A measurement of probable discharge by various types of plumbing fixtures into a drainage system.

To properly size horizontal drains, building drains and vent systems, a total dfu value must be established for each of these.

IPC 2003, Table 709.1: This table displays the dfu value and minimum trap size for each type of fixture listed.

Example: What is the total dfu value for 2 (two) floor drains?

a. 1

b. 2

c. 3

d. 4

Ans: d

2 floor drains × 2 dfu = 4

Example: What is the total dfu value for 3 (three) automatic clothes washers in a self-service laundry?

a. 3

b. 6

c. 9

d. 12

Ans: c

3 clothes washers × 3 dfu = 9

IPC 2003, Table 709.2: In some circumstances, a specific fixture may not be included in Table 709.1. If this occurs, a dfu value can still be determined by the trap or drain size of the fixture. Table 709.2 assigns a dfu value for unlisted fixtures when either the drain of trap size is known.

Example: What is the drainage fixture unit value for a dishwasher in a restaurant on a 2 inch drain?

a. 2

b. 3

c. 4

d. 5

Ans: b

Building Drains

Building Drain: The lowest part of a drainage system that receives discharge from soil, waste and other drainage pipes inside and delivers the waste to the building sewer.

3 Types of Building Drains

Sanitary: tranports sewage only
Storm: transports storm drainage
Combined: transports storm drainage and sewage

A building drain is sized according to the tallied drainage fixture unit value conveyed through the drainage system.

IPC 2003, Table 710.1(1): This table limits the number of dfus connected to any portion of the building drain or the building sewer according to the slope per foot.

Example: A building drain having ¼" slope receives a total discharge of 200 dfus. What size drain is required to convey this amount of waste?

a. 2½"

b. 3"

c. 4"

d. 5" **Ans: c**

Example: A 5" building drain having a ½" slope is capable of conveying a maximum discharge of how many dfus?

a. 480

b. 575

c. 840

d. 1,000 **Ans: b**

Notes

Notes

Horizontal Fixture Branches and Stacks

Horizontal Branch Drain: Drainage pipe extending laterally from a soil or waste stack or building drain that receives discharge from 2 or more fixture drains and transports the discharge to the soil, waste stack or building drain.

Stack: Any vertical line of soil or waste that extends through at least story.

IPC 2003, Table 710.1(2): This table properly sizes horizontal drains and stacks according to the number of dfus being transported through them.

Column 1: Diameter of pipe
Column 2: Lists the maximum number of fixture units that may discharge into any size horizontal drain
Column 3:
Column 4:
Column 5: } sizes the stack according to the number of branch intervals

Example: What is the maximum number of fixture units allowed to discharge in a 4 inch horizontal branch?

a. 90

b. 160

c. 200

d. 360

Ans: b (Column 1 & 2)

Example: A horizontal branch that receives 500 dfu's of discharge is required to be a minimum of ______ inches.

a. 4

b. 5

c. 6

d. 8

Ans: c (Column 1 & 2)

Example: What is the maximum number of fixture units allowed to discharge in a 5 inch stack with 2 branch intervals?

a. 200

b. 240

c. 500

d. 540

Ans: d (Column 1 & 4)

IPC PRACTICE EXAM #4

Notes

The following questions are based on Chapter 7 of the International Plumbing Code book and cover drainage fixture unit values, horizontal, drains and building drains.

1. A water closet flushing 1.6 gpm and a double lavatory are connected to a horizontal branch in a dwelling unit. What is the fixture unit total that will be discharging into this branch?

   a. 4 dfu

   b. 5 dfu

   c. 6 dfu

   d. 8 dfu

2. What size horizontal branch is required to convey the discharge to the building drain from question #1?

   a. 1½"

   b. 2"

   c. 2½"

   d. 3"

3. A clothes washer, two compartment laundry tray, and a floor drain are connected to a horizontal branch in a home. What size pipe diameter is required for drainage?

   a. 1½"

   b. 2"

   c. 2½"

   d. 3"

Notes

4. A four story motel contains 6 sleeping rooms on each floor. All sleeping rooms include a bathroom group with a water closet flushing greater than 1.6 gpm. What is the total dfu value discharging into each branch interval?

   a. 30

   b. 36

   c. 120

   d. 144

5. An 8 inch stack is conveying 2,800 fixture units worth of drainage into the building drain. The building has five floors. How many dfus are emptying into this stack at each floor?

   a. 400

   b. 2,000

   c. 2,200

   d. 3,600

6. A ⅛ inch sloped building drain receives discharge from two branch intervals. Each branch interval conveys 120 fixture units worth of waste. What size building drain is required to transport the drainage to the sewer?

   a. 3"

   b. 4"

   c. 5"

   d. 6"

7. What is the maximum number of fixture units allowed to convey through a six inch building drain on a ½" slope?

   a. 575 dfu

   b. 840 dfu

   c. 1,000 dfu

   d. 2,300 dfu

8. Given: Horizontal branch #1 = 46 dfu
   Horizontal branch #2 = 38 dfu
   Horizontal branch #3 = 54 dfu

   Building drain's slope = ¼"

   What size building drain is required to convey the discharge from the horizontal branches listed above?

   a. 3"

   b. 4"

   c. 5"

   d. 6"

Notes

Notes

IPC PRACTICE EXAM #5

The following questions are based on Chapters 8 and 9 of the International Plumbing Code book and cover indirect waste and vents.

1. All horizontal indirect waste pipe exceeding ________ feet must be trapped.

   a. 2

   b. 4

   c. 6

   d. 8

2. Standpipes shall extend a maximum of ________ inches above the trap weir.

   a. 12

   b. 18

   c. 36

   d. 42

3. A vent stack is required for every drainage stack that exceeds ________________.

   a. 2 branch intervals

   b. 3 branch intervals

   c. 4 branch intervals

   d. 5 branch intervals

4. The developed length of a relief vent is measured from the farthest point of vent connection to the drainage system, then, to the point of connection to the ____________.

   a. vent stack

   b. stack vent

   c. termination outside a building

   d. All of the above

5. A sump pump, other than pneumatic, has a discharge capacity of 60 gpm. What diameter of vent pipe is required if the developed length is 78 feet?

a. 1¼"

b. 1½"

c. 2"

d. 2½"

6. Where fixture drains connect at different levels and share a common vent, which fixture is not permitted at the uppermost level?

a. Water closet

b. Urinal

c. Three compartment sink

d. Whirlpool spa

7. Air admittance valves are required to be installed how far above insulation materials?

a. 6"

b. 12"

c. 18"

d. 24"

8. A dry vent shall rise to a minimum of ______ above the flood level rim of the highest trap.

a. 4"

b. 6"

c. 8"

d. 10"

Notes

Notes

9. In order for any harmful chemicals to be discharged into a plumbing system, what must first happen to prevent injury to the drain or sewer?

    a. Chemicals must be diluted
    b. Chemicals must be neutralized
    c. Both A and B
    d. None of the above

10. Sheet copper for vent flashings must weigh a minimum of ______ oz/sq. ft.

    a. 0.5" WC
    b. 1" WC
    c. 2" WC
    d. None of the above

11. A vent that connects to a horizontal drainage branch and vents a minimum of two traps up to a maximum of eight is called a ______________.

    a. common vent
    b. circuit vent
    c. branch vent
    d. individual vent

12. What is the maximum developed length of a 2½" vent?

    a. 230'
    b. 270'
    c. 300'
    d. 360'

Notes

13. An individual vent serving as a common vent is permitted to vent _______ traps.

 a. 2

 b. 3

 c. 4

 d. 6

14. A 20 foot branch vent is 2 inches in diameter. What size branch vent would be required if the developed length is increased to 50 feet?

 a. 2"

 b. 3"

 c. 4"

 d. 5"

15. What is the maximum distance from the trap weir to the vent if the drain size is 1¼"?

 a. 3½'

 b. 5'

 c. 6'

 d. 10'

16. What is the maximum temperature of water that may be discharged into the drainage system?

 a. 120°F

 b. 140°F

 c. 160°F

 d. 180°F

Notes

17. What is the minimum size wet vent required if the fixture unit discharge to the wet vent is 5 dfus?

    a. 1½"

    b. 2"

    c. 2½"

    d. 3"

18. What size relief vent is required if connected to a 4 inch vent stack?

    a. 1"

    b. 2"

    c. 3"

    d. None of the above

19. An air admittance valve is used to vent fixtures ____________.

    a. once every 2 floors

    b. once every 3 floors

    c. on the same floor level

    d. None of the above

20. What is the maximum wet vent size permitted?

    a. 1½"

    b. 2"

    c. 2½"

    d. 3"

21. What is the minimum size air gap required for a 1½" indirect waste pipe?

    a. 1½"

    b. 2"

    c. 2½"

    d. 3"

22. Vent terminals shall terminate a minimum of ________ feet from a lot line.

    a. 10
    b. 15
    c. 20
    d. 25

23. Sheet copper for vent flashings must weigh a minimum of ________ ounces per square foot.

    a. 4
    b. 6
    c. 8
    d. 10

24. What is the maximum distance allowed from a fixture trap to the vent if the trap size is 1½" on ¼" slope?

    a. 3½'
    b. 5'
    c. 6'
    d. All of the above

25. What is the minimum size stack vent allowed to serve a drain?

    a. 1¼"
    b. 1½"
    c. 2"
    d. 2½"

Notes

Notes

IPC PRACTICE EXAM #6

The following questions are based on Chapters 10 and 11 of the International Plumbing Code book and cover traps, interceptors, separators, and storm drainage.

1. What is the maximum size liquid seal required for fixture traps?

   a. 1½"

   b. 2"

   c. 3"

   d. 3"

2. A cast-iron subsoil drain pipe must conform to all of the following standards except ________.

   a. ASTM B 74

   b. ASTM A 888

   c. CISPI 301

   d. ASTM A 74

3. What is the minimum slope required for the horizontal branches of a building storm drain?

   a. ⅛ unit vertical in 12 units horizontal

   b. ¼ unit vertical in 12 units horizontal

   c. ½ unit vertical in 12 units

   d. ¾ unit vertical in 12 units horizontal

4. A 12,000 square foot roof has six leaders installed for storm drainage. How many square feet of roof area is each leader responsible for draining?

   a. 6,000 sq. ft.

   b. 4,000 sq. ft.

   c. 2,000 sq. ft.

   d. Not enough information

5. A sump pit shall be constructed from what type of material?

   a. Plastic

   b. Steel

   c. Tile

   d. All of the above

6. Discharge piping from an ejector in a dwelling unit is required to have what kind of valve?

   a. Check valve

   b. Gate valve

   c. Globe valve

   d. No valve is required

7. Which of the following is a prohibited trap?

   a. Bell trap

   b. Crown-vented trap

   c. Drum trap

   d. All of the above

8. Which of the following is permitted to discharge storm water out onto a street?

   a. A drug store having at least 25' of clearance in front of building

   b. A ten unit apartment complex

   c. A convenience store surrounded by a concrete parking lot

   d. A two-family dwelling

Notes

Notes

9. Given: 1 inch per hour rainfall rate

   Two gpm of continuous discharge into building storm drain is equivalent to ________ of roof area.

   a. 96 sq. ft.
   b. 192 sq. ft.
   c. 288 sq. ft.
   d. 384 sq. ft.

10. Roof drains have strainers which must extend a minimum of ________ inches above the roof's surface.

    a. 1½
    b. 2
    c. 3
    d. 4

11. What is the maximum distance allowed from a fixture outlet to a trap weir?

    a. 12"
    b. 18"
    c. 24"
    d. 36"

12. What is the retention capacity of a grease trap with a 7 gpm flow through rating?

    a. 14 lbs.
    b. 18 lbs.
    c. 20 lbs.
    d. 24 lbs.

13. Given: 7 inches rainfall per hour

What is the maximum square feet of roof area a single 4 inch leader can convey?

a. 1,260 sq. ft.

b. 2,630 sq. ft.

c. 3,070 sq. ft.

d. 4,945 sq. ft.

14. What is the minimum diameter of a subsoil drain?

a. 2"

b. 3"

c. 4"

d. 6"

15. A sump pit's discharge piping must include a ____________.

a. full flow check valve

b. double check valve

c. gate valve

d. gate valve and full flow check valve

16. What is the minimum number of roof drains required to be installed on a 10,000 square foot roof area?

a. 2

b. 4

c. 5

d. 10

Notes

Notes

17. Subsoil drains shall discharge to a ______________.

    a. sump
    b. dry well
    c. approved location above ground
    d. All of the above

18. Which of the following is not required to have an oil separator?

    a. A mechanic's garage
    b. Commercial parking garage storing automobiles
    c. A parking garage which dispenses gasoline
    d. Car wash

19. Slip joints shall be installed only ______________.

    a. on the trap inlet
    b. on the trap outlet
    c. within the trap seal
    d. All of the above

20. Sand interceptors shall have a water seal not less than ________.

    a. 6"
    b. 12"
    c. 18"
    d. 24"

21. What size trap is required for a 4 inch storm drain connected to a combined sewer?

    a. 3"
    b. 4"
    c. 5"
    d. None of the above

22. What is the minimum size water seal required of an outlet opening to an oil separator?

    a. 6"

    b. 12"

    c. 18"

    d. 24"

23. Given: 4" rainfall per hour
    ⅛" slope

    What size horizontal storm drain is required for 5,500 sq. ft. of roof drainage?

    a. 4"

    b. 5"

    c. 6"

    d. 8"

24. A 15,000 square foot roof area is required to have at least how many roof drains?

    a. 2

    b. 4

    c. 5

    d. 10

25. Given: 5" rainfall per hour
    ½" slope

    What is the maximum square footage a 3 inch horizontal storm drain can convey?

    a. 657 sq. ft.

    b. 928 sq. ft

    c. 1,310 sq. ft

    d. 3,010 sq. ft.

Notes

Notes

STORM DRAIN SIZING

This section primarily addresses Tables 1106.2 and 1106.3 in Chapter 11 of the International Plumbing Code book.

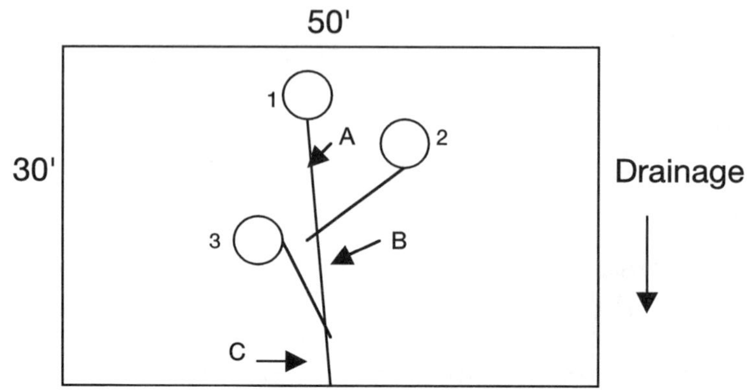

3" rainfall per hour
⅛" slope

○ = leaders
____ = horizontal storm drain piping

Leader: An exterior drainage pipe for conveying storm water from roof or gutter drains to an approved means of disposal.

I. Sizing the Leaders: Table 1106.2

1. Find the square footage of the roof
   30' × 50' = **1500 sq. ft.**

2. Divide the square footage of the roof by the number of leaders displayed in the above diagram.
   1500 sq. ft. ÷ 3 leaders = **500**

This means that each leader is responsible for conveying 500 sq. ft. worth of storm water drainage.
Using *Table 1106.2*, properly size the leaders according to code.

Rainfall rate: 3" per hour (find this rate at the top of the table) The numerical values below this column is the projected square footage of drainage. We have already predetermined that each leader will be conveying 500 sq. ft. of storm water.

Find the numerical value that will accommodate at least 500 sq. ft. Once identified, follow that row to the left side of Table1106.2 to determine the diameter of each leader.

Answer: **2"**

II. Sizing Horizontal Storm Drain Piping: Table 1106.3

Notes

Section A:

1. When sizing horizontal storm drain piping, the total square footage conveying through each section must be determined.

For example: How many square feet of drainage is flowing through Section A? **500 sq. ft.**

Using *Table 1106.3*, find the 3" rainfall rate column and the ⅛" slope. Locate the numerical values under the 3" column and find the one that will accommodate at least 500 sq. ft. of drainage. Once identified, follow that row to the left side of the table to determine the correct size of Section A horizontal piping.

Answer: **3"**

Section B:

1. Conveying 500 sq. ft. of drainage from Leader 1 and 500 sq. ft. from Leader 2. Total drainage is 1,000 sq. ft.

Answer: **3"**

Section C:

1. Conveying 500 sq. ft. of drainage from Leader 1, 500 sq. ft. from Leader 2, and 500 sq. ft. from Leader 3. Total drainage is 1,500 sq. ft.

Answer: **4"**

Notes

STORM DRAIN DIAGRAM

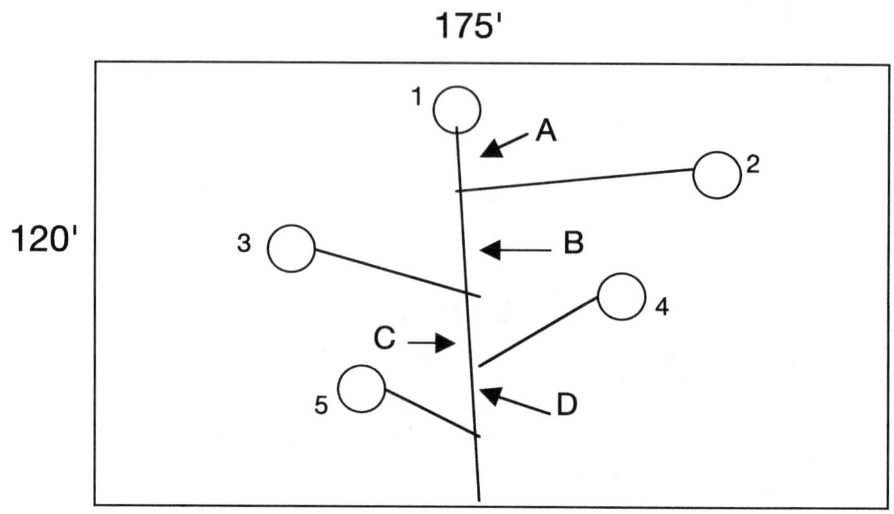

5" Rainfall rate per hour
$^{1}/_{4}$" slope

Leader Size: ______________________inches

Horizontal A: ______________________inches

Horizontal B: ______________________inches

Horizontal C: ______________________inches

Horizontal D: ______________________inches

PLUMBING ISOMETRICS

The following exercise, based on Chapters 7 and 9 of the IPC, will demonstrate how to properly size building drains, horizontal drains and vents

Figure 3

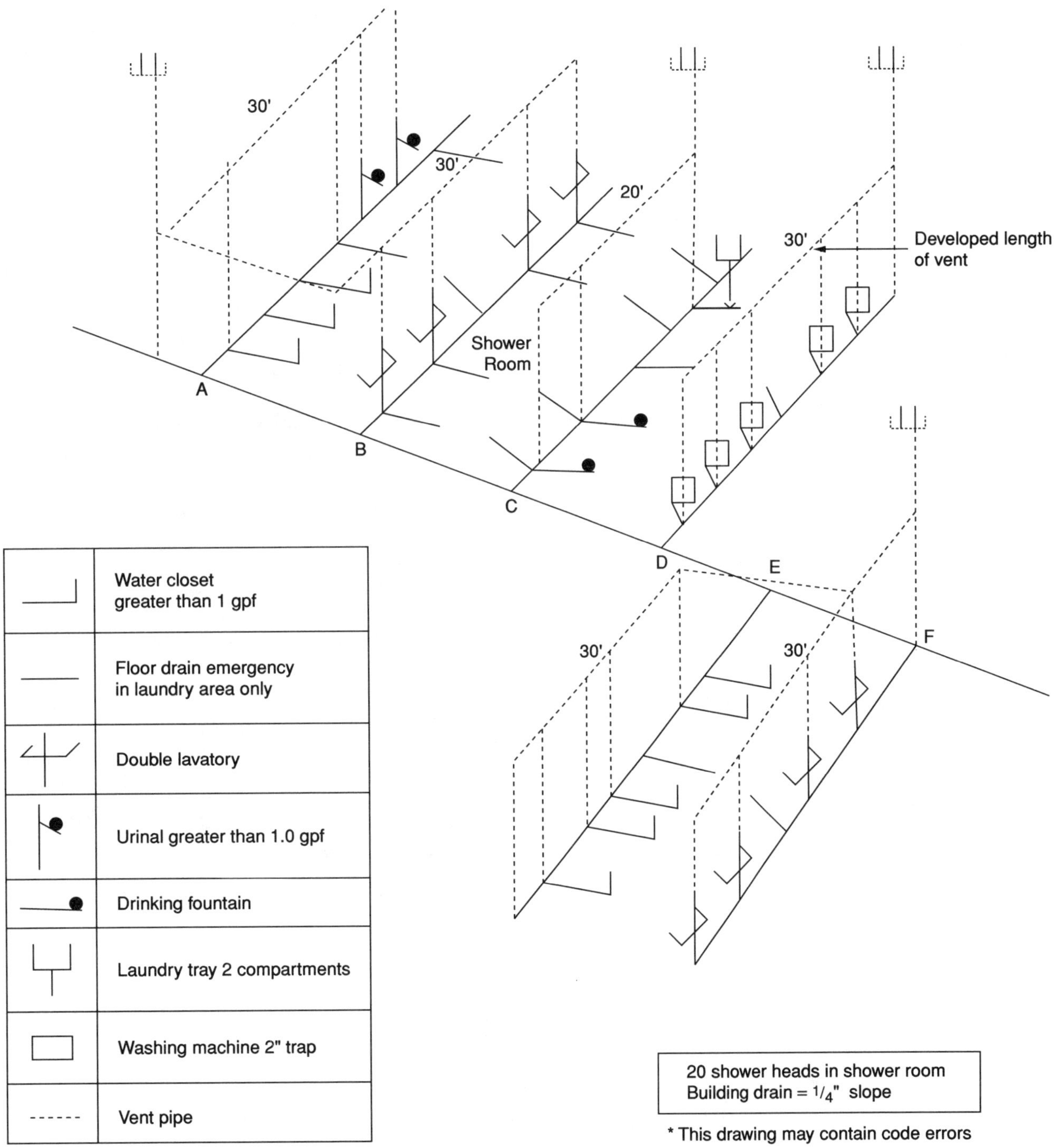

Notes

Drainage Fixture Units (Figure 3) Use Table 709.1 for dfu Value

	# Fixtures		dfu Value		Total
Water Closets	______	×	______	=	______
Urinals	______	×	______	=	______
Lavatories	______	×	______	=	______
Floor drains	______	×	______	=	______
Shower heads	______	×	______	=	______
Water fountains	______	×	______	=	______
Washing Machines	______	×	______	=	______
Laundry tray	______	×	______	=	______
			Total	=	______

Horizontal Drain Sizing (Figure 3)

Tip: When calculating a drainage fixture unit total for a horizontal drain, make a list of each type of fixture on the branch. This will help ensure all fixture types are accounted for. From this point, an accurate tally of drainage fixture units can be determined to correctly size the horizontal drain.

	Fixture Type	# of		dfu Value	Total
Horizontal Branch A:	Water Closet	3	×	6	18
	Floor Drains	2	×	2	4
	Urinals	2	×	4	8
					= 30

Horizontal branch A has a total of 30 dfus. Using **Table 710.1(2)** an accurate drain size can be determined.

What is the minimum drain size for horizontal branch A? ***Ans:*** *4"*

Use the example on page 89 to successfully complete the rest of the worksheet.

	Fixture Type	# of		dfu Value		Total
Horizontal Branch B:	______	____	×	______	=	____
	______	____	×	______	=	____
	______	____	×	______	=	____
				Total	=	____

	Fixture Type	# of		dfu Value		Total
Horizontal Branch C:	______	____	×	______	=	____
	______	____	×	______	=	____
	______	____	×	______	=	____
	______	____	×	______	=	____
				Total	=	____

	Fixture Type	# of		dfu Value		Total
Horizontal Branch D:	______	____	×	______	=	____
	______	____	×	______	=	____
				Total	=	____
Horizontal Branch E:	______	____	×	______	=	____
	______	____	×	______	=	____
				Total	=	____
Horizontal Branch F:	______	____	×	______	=	____
	______	____	×	______	=	____
				Total	=	____

Each branch dfu total can now be used to size the horizontal drains.

1. What is the minimum drain size for horizontal branch B? ______
2. What is the minimum drain size for horizontal branch C? ______
3. What is the minimum drain size for horizontal branch D? ______
4. What is the minimum drain size for horizontal branch E? ______
5. What is the minimum drain size for horizontal branch F? ______

Notes

Notes

Sizing Building Drain (Figure 3)

When sizing a specific section of a building drain, one must first determine the total number of drainage fixture units flowing into that particular zone of pipe. Using Table 710.1(1), choose the correct column according to the given slope and size that specified section of drain pipe.

Building Drain A–B: This zone is conveying all fixture units discharged from horizontal branch A: (30 dfus). The slope, given in figure 3, is ¼". Column 4 of Table 710.1(1) determines the maximum number of dfus allowed to connect to a certain diameter of pipe having a ¼" slope.

	Total dfu Value		Building Drain Size
Building Drain A–B:	30	=	3"
Building Drain B–C:	_____	=	_____
Building Drain C–D:	_____	=	_____
Building Drain D–E:	_____	=	_____
Building Drain E–F:	_____	=	_____

Vent Sizing (Figure 3)

IPC 2003, Table 916.1: This is the most common table used for sizing vents. The drain size and the number of fixture units being vented must be known in order to properly size the vent.

If you notice, in the first column, there are multiple listings of the same diameter of pipe. To ensure the correct diameter is chosen, the total fixture units being vented must be selected from column 2. The next step is to determine the developed length of vent pipe, which is usually given in the figure illustrated or the question. Everything needed to determine the vent size is in Table 916.1. Make sure you pay close attention to what each column represents.

	Drain Size	# FU Vented	Developed Length		Vent Size
Horizontal Branch A:	3"	30	30'	=	2"
Horizontal Branch B:	_____	_____	_____	=	_____
Horizontal Branch C:	_____	_____	_____	=	_____
Horizontal Branch D:	_____	_____	_____	=	_____
Horizontal Branch E:	_____	_____	_____	=	_____
Horizontal Branch F:	_____	_____	_____	=	_____

ISOMETRIC PRACTICE EXAM

Note: Drawings not to scale

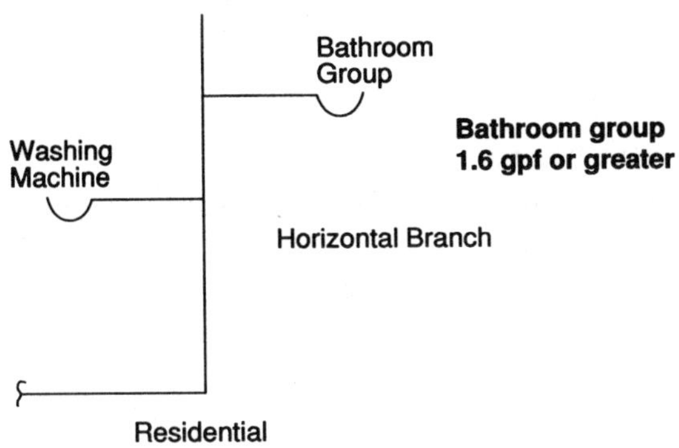

Refer to the drawing above.

1. What is the total drainage fixture units?

   a. 4 dfu

   b. 6 dfu

   c. 10 dfu

   d. 8 dfu

2. What size should the horizontal branch be?

   a. 1½"

   b. 3"

   c. 4"

   d. 2½"

Notes

Notes

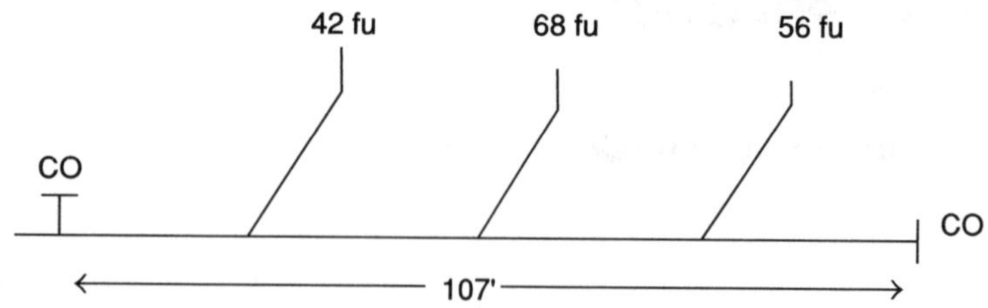

Refer to the drawing above.

1. What size building drain would be required to accommodate the total dfus?

   a. 4"

   b. 3"

   c. 2"

   d. 6"

2. What is wrong in this drawing as per most codes?

   a. 107' is more than maximum distance between clean outs

   b. There are too many branch lines

   c. Nothing is wrong

   d. Drainage fixture units are not the same in each branch

3. What size should building drain be with ⅛ slope?

   a. 2"

   b. 3"

   c. 4"

   d. 6"

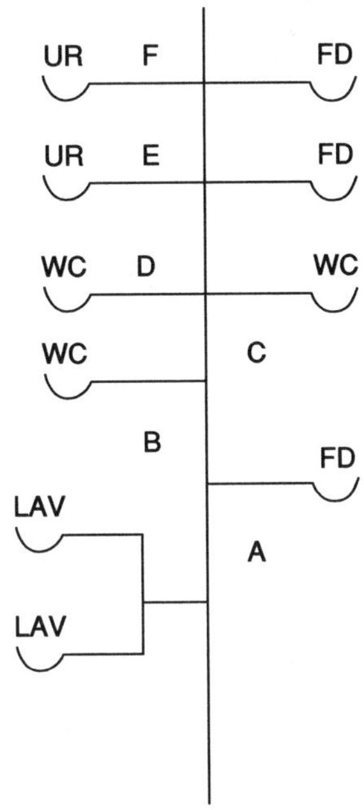

Refer to the drawing above.

1. What is the total drainage fixture units if WC's are tank type?

 a. 26 dfu

 b. 10 dfu

 c. 18 dfu

 d. 28 dfu

2. At what connection would branch drain size have to be increased from 3" to 4"?

 a. C

 b. A

 c. F

 d. D

Notes

Notes

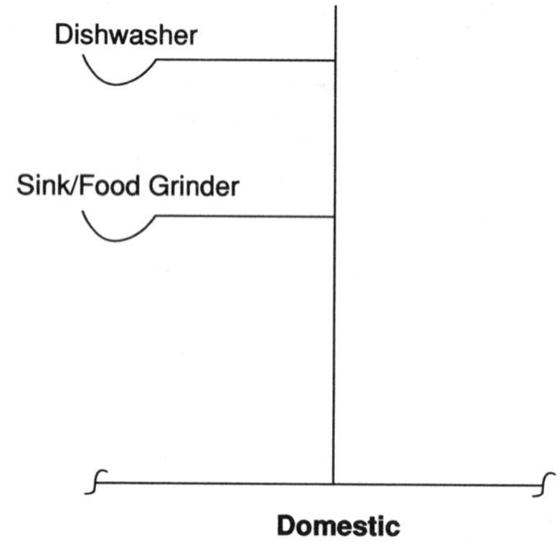

Refer to the drawing above.

1. What should be the size of the dishwasher drain line?

   a. 2"
   b. 3"
   c. 1½"
   d. 1¼"

2. What should be the size of the sink/food grinder line?

   a. 1¼"
   b. 1½"
   c. 2"
   d. 3"

3. What would be the total drain fixture units (dfu) for the vertical drain stack?

   a. 5 dfu
   b. 2 dfu
   c. 6 dfu
   d. 4 dfu

4. What should be the size of the vertical drain stack?

   a. 1½"

   b. 3"

   c. 2"

   d. 4"

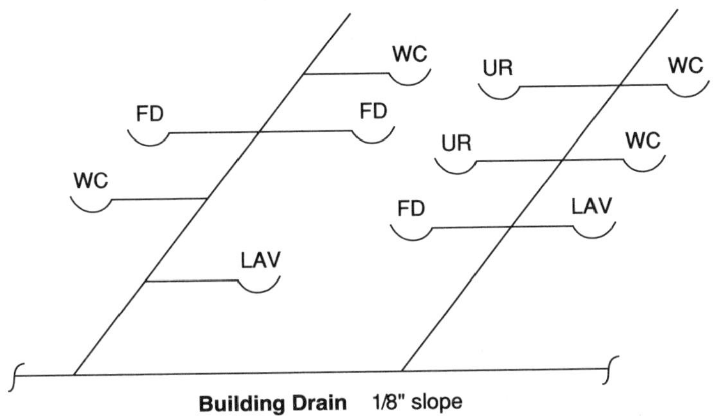

Refer to the drawing above.

1. What is the total drainage fixture units (public use 1.6 gpf or greater)?

   a. 17 dfu

   b. 40 dfu

   c. 23 dfu

   d. 36 dfu

2. What is the size of branch A?

   a. 3"

   b. 4"

   c. 6"

   d. 2"

Notes

Notes

3. What is the size of branch B?

   a. 6"

   b. 3"

   c. 4"

   d. 2"

4. What size should the building drain be?

   a. 3"

   b. 4"

   c. 6"

   d. 8"

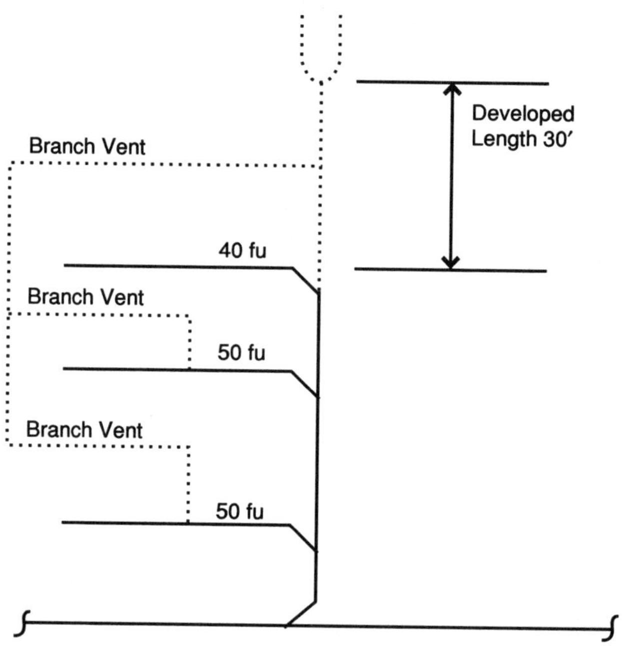

Refer to the drawing above

1. What size drain line for horizontal branch C?

   a. 2"

   b. 3"

   c. 4"

   d. 6"

2. What size drain line for horizontal branch B and A?

 a. 6"
 b. 4"
 c. 3"
 d. 2"

3. What size drain line for the vertical drain stack?

 a. 2"
 b. 6"
 c. 3"
 d. 4"

4. What size should the main stack be?

 a. 4"
 b. 6"
 c. 3"
 d. 2"

5. What size should branch vent from horizontal branch A be?

 a. 1¼"
 b. 1½"
 c. 2"
 d. 2½"

6. What size should branch vent from horizontal branch B be?

 a. 1½"
 b. 1¼"
 c. 2½"
 d. 2"

Notes

Notes

7. What size should branch vent from horizontal branch C be?

   a. 1¼"

   b. 2"

   c. 1½"

   d. 2½"

8. What size should the main vent stack be?

   a. 1½"

   b. 2"

   c. 2½"

   d. 3"

Use the following drawings and tables for this practice test.

- Water closets in the locker rooms are flush valve type, and the water closets in kitchen bathrooms are tank type.
- Assume all water closets are 'public' at 1.6 gpf or greater.
- Urinals are wall hung.
- The dishwasher is installed on a 2" trap and does not increase fixture load.
- The A/C condensate line at C12 is rated at 1 fixture unit.
- The five waste outlets in each shower room serve two shower heads each. Count each waste outlet in each shower room as 2 domestic showers.
- Unless otherwise stated, all drain lines are installed on a slope of ¼" per foot, all vents are less than 40 ft. long.
- Drainage plumbing under slab.

Notes

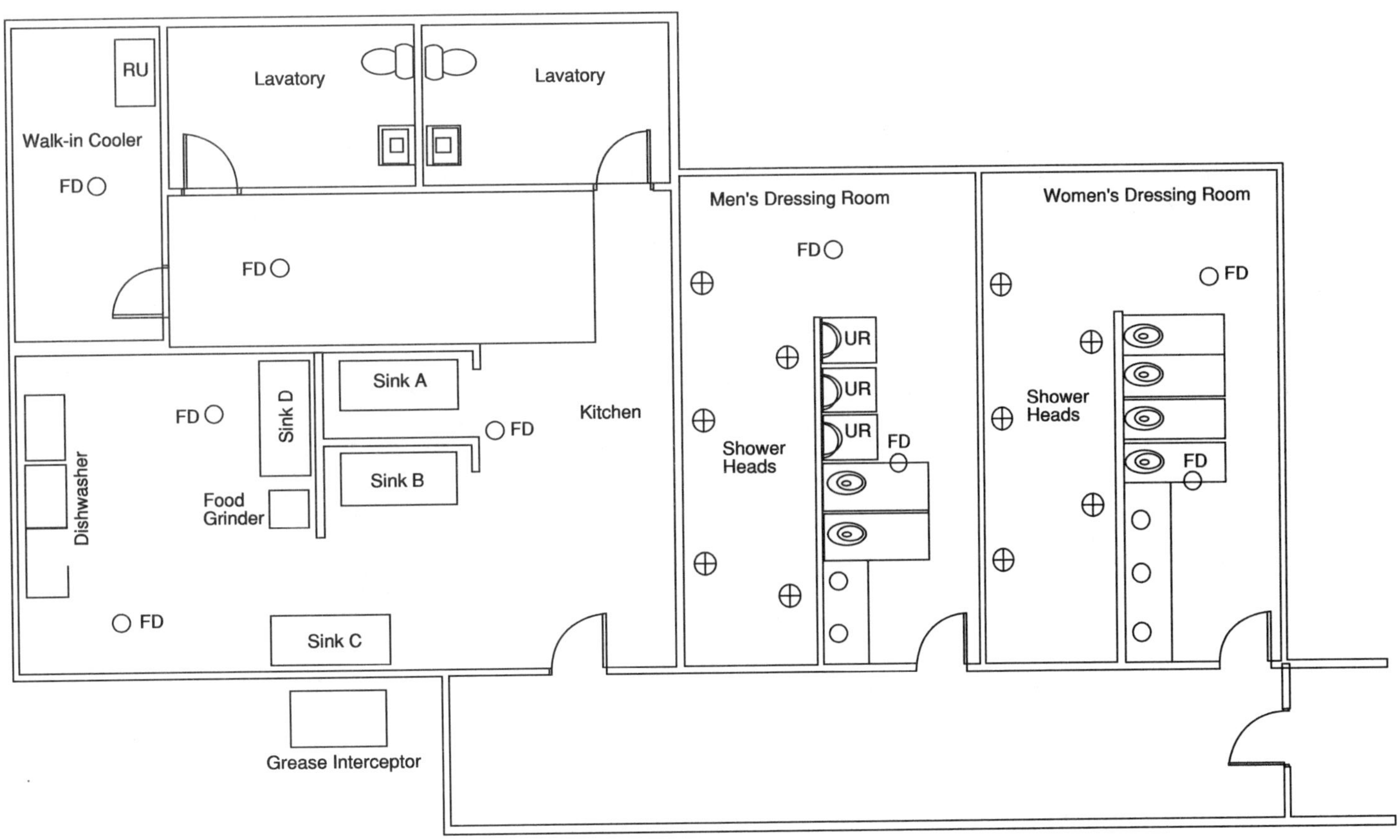

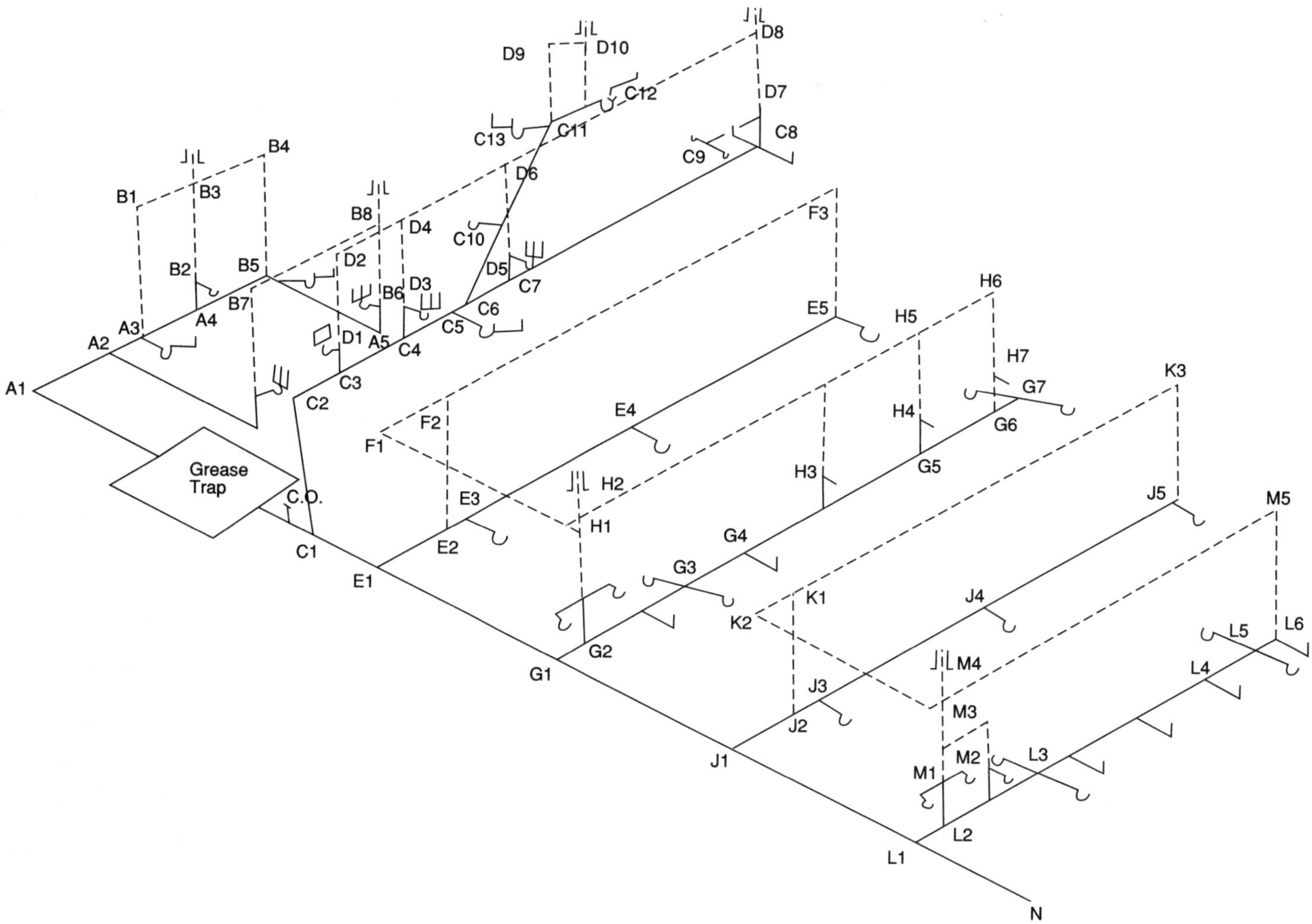
A1
A2
A3
A4
A5
B1
B2
B3
B4
B5
B6
B7
B8
Grease
Trap
C.O.
C1
C2
C3
C4
C5
C6
C7
C8
C9
C10
C11
C12
C13
D1
D2
D3
D4
D5
D6
D7
D8
D9
D10
E1
E2
E3
E4
E5
F1
F2
F3
G1
G2
G3
G4
G5
G6
G7
H1
H2
H3
H4
H5
H6
H7
J1
J2
J3
J4
J5
K1
K2
K3
L1
L2
L3
L4
L5
L6
M1
M2
M3
M4
M5
N

TABLE 709.1
DRAINAGE FIXTURE UNITS FOR FIXTURES AND GROUPS

Fixture type	Drainage fixture unit value as load factors	Minimum size of trap (inches)
Automatic clothes washers, commercial[a]	3	2
Automatic clothes washers, residential	2	2
Bathroom group consisting of water closet, lavatory, bidet and bathtub or shower	6	—
Bathtub[b] (with or without overhead shower or whirlpool attachments)	2	1½
Bidet	2	1¼
Combination sink and tray	2	1½
Dental lavatory	1	1¼
Dental unit or cuspidor	1	1¼
Dishwashing machine,[c] domestic	2	1½
Drinking fountain	½	1¼
Emergency floor drain	0	2
Floor drains	2	2
Kitchen sink, domestic	2	1½
Kitchen sink, domestic with food waste grinder and/or dishwasher	2	1½
Laundry tray (1 or 2 compartments)	2	1½
Lavatory	1	1¼
Shower compartment, domestic	2	2
Sink	2	1½
Urinal	4	Footnote d
Urinal, 1 gallon per flush or less	2[e]	Footnote d
Wash sink (circular or multiple) each set of faucets	2	1½
Water closet, flushometer tank, public or private	4[e]	Footnote d
Water closet, private installation	4	Footnote d
Water closet, public installation	6	Footnote d

For SI: 1 inch = 25.4 mm, 1 gallon = 3.785 L.
[a]For traps larger than 3 inches, use Table 709.2.
[b]A showerhead over a bathtub or whirlpool bathtub attachments does not increase the drainage fixture unit value.
[c]See Sections 709.2 through 709.4 for methods of computing unit value of fixtures not listed in Table 709.1 or for rating of devices with intermittent flows.
[d]Trap size shall be consistent with the fixture outlet size.
[e]For the purpose of computing loads on building drains and sewers, water closets or urinals shall not be rated at a lower drainage fixture unit unless the lower values are confirmed by testing.

TABLE 709.2
DRAINAGE FIXTURE UNITS FOR FIXTURE DRAINS OR TRAPS

Fixture drain or trap size (inches)	Drainage fixture unit value
1¼	1
1½	2
2	3
2½	4
3	5
4	6

For SI: 1 inch = 25.4 mm.

TABLE 710.1(1)
BUILDING DRAINS AND SEWERS

Diameter of pipe (inches)	Maximum number of drainage fixture units connected to any portion of the building drain or the building sewer, including branches of the building drain[a]			
	Slope per foot			
	1/16 inch	⅛ inch	¼ inch	½ inch
1¼	—	—	1	1
1½	—	—	3	3
2	—	—	21	26
2½	—	—	24	31
3	—	36	42	50
4	—	180	216	250
5	—	390	480	575
6	—	700	840	1,000
8	1,400	1,600	1,920	2,300
10	2,500	2,900	3,500	4,200
12	2,900	4,600	5,600	6,700
15	7,000	8,300	10,000	12,000

For SI: 1 inch = 25.4 mm, 1 inch per foot = 0.0833 mm/in.
[a]The minimum size of any building drain serving a water closet shall be 3 inches.

TABLE 710.1(2)
HORIZONTAL FIXTURE BRANCHES AND STACKS[a]

Diameter of pipe (inches)	Maximum number of drainage fixture units (dfu)			
		Stacks[b]		
	Total for a horizontal branch	Total discharge into one branch interval	Total for stack of three branch intervals or loss	Total for stack greater than three branch intervals
1½	3	2	4	8
2	6	6	10	24
2½	12	9	20	42
3	20	20	48	72
4	160	90	240	500
5	360	200	540	1,100
6	620	350	960	1,900
8	1,400	600	2,200	3,600
10	2,500	1,000	3,800	5,600
12	3,900	1,500	6,000	8,400
15	7,000	Footnote c	Footnote c	Footnote c

For SI: 1 inch = 25.4 mm.

[a]Does not include branches of the building drain. Refer to Table 710.1(1).
[b]Stacks shall be sized based on the total accumulated connected load at each story or branch interval. As the total accumulated connected load decreases, stacks are permitted to be reduced in size. Stack diameters shall not be reduced to less than one-half of the diameter of the largest stack size required.
[c]Sizing load based on design criteria.

TABLE 916.1
SIZE AND DEVELOPED LENGTH OF STACK VENTS AND VENT STACKS

Diameter of soil or waste stack (inches)	Total fixture units being vented (dfu)	Maximum developed length of vent (feet)[a] Diameter of vent (inches)										
		1¼	1½	2	2½	3	4	5	6	8	10	12
1¼	2	30										
1½	8	50	150									
1½	10	30	100									
2	12	30	75	200								
2	20	26	50	150								
2½	42		30	100	300							
3	10		42	150	360	1,040						
3	21		32	110	270	810						
3	53		27	94	230	680						
3	102		25	86	210	620						
4	43			35	85	250	980					
4	140			27	65	200	750					
4	320			23	55	170	640					
4	540			21	50	150	580					
5	190				28	82	320	990				
5	490				21	63	250	760				
5	940				18	53	210	670				
5	1,400				16	49	190	590				
6	500					33	130	400	1,000			
6	1,100					26	100	310	780			
6	2,000					22	84	260	660			
6	2,900					20	77	240	600			
8	1,800						31	95	240	940		
8	3,400						24	73	190	720		
8	5,600						20	62	160	610		
8	7,600						18	56	140	560		
10	4,000							31	78	310	960	
10	7,200							24	60	240	740	
10	11,000							20	51	200	630	
10	15,000							18	46	180	570	
12	7,300								31	120	380	940
12	13,000								24	94	300	720
12	20,000								20	79	250	610
12	26,000								18	72	230	500
15	15,000									40	130	310
15	25,000									31	96	240
15	38,000									26	81	200
15	50,000									24	74	180

For SI: 1 inch = 25.4 mm, 1 foot = 304.8 mm.
[a]The developed length shall be measured from the vent connection to the open air.

Practice Test for Fixture Units

1. How many drainage fixture units (dfu) at A1?

   a. 8
   b. 10
   c. 12
   d. 14

2. What is the drainage pipe size required for branch A1?

   a. 1½"
   b. 3"
   c. 4"
   d. 2"

3. What is the drainage fixture unit (dfu) count for C1?

   a. 25
   b. 20
   c. 22
   d. 32

4. What size drainage pipe is required between C9 and D7?

   a. 1¼"
   b. 2½"
   c. 2"
   d. 1½"

Notes

Notes

5. What type of fitting at C9? Fitting supports 2 lavs?

   a. Wye
   b. Sanitary tee
   c. Sanitary cross
   d. Double 90° elbow

6. How many dfu for branch G1?

   a. 25
   b. 38
   c. 42
   d. 32

7. What size should the cleanout by the grease trap be?

   a. 2"
   b. 3"
   c. 4"
   d. 2½"

8. What size would be the vent pipe at B2?

   a. 1½"
   b. 1¼"
   c. 2"
   d. 2½"

9. What is the dfu count at J1?

   a. 3
   b. 6
   c. 15
   d. 12

10. How many dfu for branch L1?

    a. 41
    b. 39
    c. 34
    d. 36

11. What type of fitting is required at L4?

    a. 4" wye
    b. 3" sanitary tee
    c. 3" combo y and ⅛" bend
    d. 3" elbow

12. What is dfu count for vent thru roof at B3?

    a. 6
    b. 4
    c. 8
    d. 10

13. What type of fitting between trap arm M1 and M2 fitting supports 2 lavs?

    a. 1½" cross
    b. 2" cross
    c. 1½" double sanitary tee
    d. 2" × 1½" double sanitary tee

14. What is the size of the building drain?

    a. 3"
    b. 6"
    c. 4"
    d. 8"

Notes

Notes

15. What would be the size of the vent through the roof at M4 if the developed length is 35 feet?

    a. 2"

    b. 2½"

    c. 3"

    d. 4"

16. What is the trap size at E5 if each shower drain has 2 shower heads?

    a. 1½"

    b. 3"

    c. 2½"

    d. 2"

17. What is the size of the vent through the roof at H2 if the developed length is 32 feet?

    a. 2"

    b. 2½"

    c. 1½"

    d. 3"

PART FIVE

International Plumbing Code 2003 Final Review Exam

Notes

1. A laundry tray has an effective opening of 1½ inches. What is the minimum size air gap required if close to a wall?

   a. 1.5"

   b. 2.5"

   c. 3"

   d. None of the above

2. Water hammer arresters shall conform to which of the following standards?

   a. ASSE 1010

   b. ASSE 1066

   c. ASSE 5020

   d. ASSE 5047

3. If a pump is rated 20 psi, what pressure is required to perform a forced sewer test?

   a. 5 psi

   b. 10 psi

   c. 20 psi

   d. 25 psi

4. Which of the following requires a cleanout to be installed in a building drain?

   a. 30 degree change in direction

   b. 35 degree change in direction

   c. 45 degree change in direction

   d. 90 degree change in direction

5. A water heater in an attic must have a clear access opening of ___________.

   a. 20" × 30"

   b. 22" × 30"

   c. 24" × 30"

   d. 30" × 30"

6. What is the minimum-size water-closet compartment allowed by code?

   a. 15"

   b. 21"

   c. 30"

   d. 1,800 sq. in.

7. Steel pipe shall be horizontally supported every _____ feet.

   a. 5

   b. 10

   c. 12

   d. 15

8. Which of the following is required to be a minimum distance of 25 feet away from an individual water supply?

   a. Barnyard

   b. Septic tank

   c. Sewer

   d. Pasture

Notes

Notes

9. What is the minimum trap size for an emergency floor drain?

   a. 1¼"

   b. 1½"

   c. 2"

   d. None of the above

10. A standpipe for a clothes washer must extend a minimum of _____ inches above the trap weir.

    a. 18

    b. 24

    c. 36

    d. 42

11. A theater has a maximum occupant load of 500 people. What is the minimum number of water closets required for the female patrons?

    a. 2

    b. 3

    c. 4

    d. 5

12. A removable strainer from a shower waste outlet must be at least ______ inches in diameter.

    a. 2

    b. 3

    c. 4

    d. 5

13. A minimum ½ inch fixture supply pipe is required for all of the following except?

    a. Bathtub
    b. Dishwasher
    c. One piece water closet
    d. Drinking fountain

14. What size horizontal branch is required to serve 620 drainage fixture units?

    a. 5"
    b. 6"
    c. 8"
    d. 10"

15. What is the minimum diameter of service pipe required to be installed in a residential service?

    a. ½"
    b. ¾"
    c. 1"
    d. 1½"

16. The minimum opening of a scupper to prevent the depth of ponding water from exceeding the roof's capability is ______ inches.

    a. 2
    b. 3
    c. 4
    d. 5

Notes

Notes

17. What is the maximum water pressure allowed within a building?

    a. 80 psi

    b. 100 psi

    c. 140 psi

    d. 160 psi

18. Given: Drain size = 4 inches

    Fixture units being vented = 260
    Developed length of vent = 60 feet
    What is the minimum size vent stack required?

    a. 1½"

    b. 2"

    c. 2½"

    d. 3"

19. A drainage and vent air test is performed until there is a uniform gauge pressure of 5 psi and must be maintained a minimum of ____________.

    a. 5 minutes

    b. 10 minutes

    c. 15 minutes

    d. 20 minutes

20. When backfilling, the crown of the pipe must be covered by _______ of tamped earth.

    a. 6"

    b. 12"

    c. 18"

    d. 24"

21. What is the maximum occupant load allowed before separate facilities are required?

    a. 2

    b. 4

    c. 10

    d. 15

22. What is the minimum clearance in front of a water closet to any wall?

    a. 15"

    b. 21"

    c. 30"

    d. 60"

23. What is the fixture unit value for an emergency floor drain?

    a. 0

    b. 1

    c. 2

    d. 4

24. What is the minimum size wet vent permitted to be used?

    a. 1½"

    b. 2"

    c. 2½"

    d. 3"

25. Which of the following is detrimental to the sewer system?

    a. Ashes

    b. Cinders

    c. Grease

    d. All of the above

Notes

Notes

PART SIX
International Fuel Gas Code

The following section covers the types of questions that are likely to be asked on most plumbing exams that include the *International Fuel Gas Code* as a reference.

Notes

INTERNATIONAL FUEL GAS CODE 2003 PRACTICE EXAM

1. What is the minimum burial depth required for an individual gas line connected to a grill?

    a. 8"

    b. 12"

    c. 18"

    d. 24"

2. If a 1 inch pipe needs to be field threaded, what is the approximate number of threads to be cut?

    a. 8

    b. 10

    c. 12

    d. 13

3. What is the approximate gas input for a built-in oven installed in a single family dwelling?

    a. 65,000 Btu/h

    b. 40,000 Btu/h

    c. 25,000 Btu/h

    d. 3,000 Btu/h

4. Which of the following is the standard weight for steel and wrought-iron pipe?

    a. Schedule 40

    b. Schedule 80

    c. Schedule 120

    d. All of the above

5. Which of the following materials is used to make a flange gasket?

   a. Metal-jacketed asbestos
   b. Aluminum O-ring
   c. Asbestos
   d. All of the above

6. The maximum amount of phosphorous allowed in brazing alloy is ______.

   a. 0.02%
   b. 0.04%
   c. 0.05%
   d. 0.08%

7. Which of the following appliances can have a fuel connector up to 6 feet in length?

   a. Ventless wall heater
   b. Dryer
   c. Range
   d. Both B and C

8. ¾ inch gas tubing is required to be supported every ______.

   a. 4'
   b. 6'
   c. 8'
   d. every floor level

Notes

Notes

9. Medium pressure regulators must comply with all of the following except ______________.

   a. a tee fitting with one opening capped shall be installed between the MP regulator and the upstream shut-off valve

   b. an MP regulator shall maintain a reduced outlet pressure under lock-up conditions

   c. an MP regulator must be checked for leaks once every six months or according to the manufacturer's recommendation

   d. an MP regulator shall be suitable for the inlet and outlet gas pressures

10. What is the maximum depth a joist can be notched on the ends?

    a. 20%

    b. 25%

    c. 33%

    d. 40%

11. An emergency shut-off device for a fuel dispenser shall be located at least ______ feet away dispenser.

    a. 25

    b. 50

    c. 75

    d. 100

12. Given: 0.5 psi
    0.5" WC

    A 50 foot pipe is connected to a barbeque. What size gas pipe is required?

    a. ¼"

    b. ⅜"

    c. ½"

    d. ¾"

13. What is the maximum operating pressure for a piping system located in a building?

    a. 5 psi

    b. 10 psi

    c. 15 psi

    d. 20 psi

14. In non–load bearing partitions, any wood stud is permitted to be notched, however, the maximum depth allowed shall not exceed ________.

    a. 20%

    b. 25%

    c. 40%

    d. 60%

15. The quantity of heat required to raise one pound of water one degree fahrenheit is known as a _______.

    a. demand

    b. watt

    c. Btu

    d. None of the above

Notes

Notes

GAS PIPE SIZING QUESTIONNAIRE

Step 1: Find the longest run of pipe from the point of delivery (meter) to the most remote outlet.

Once that length is determined, look at the left hand column of Table 402.4(2) and find that run of pipe. If the length falls in between two of the numbers round up to the next size.

When the correct length is found on the table, every section of pipe will be sized according to that row. All other lengths of pipe remaining are no longer important.

Step 2: Convert Btu/h into cubic feet of gas Btu ÷ 1,000

Example: *Convert 35,000 Btu to cubic feet of gas*

Ans: 35,000 ÷ 1,000 = 35 cubic feet of gas

Step 3: You are now ready to size gas pipe. Find the longest run of gas pipe and size every section according to that length. The numbers in the middle of the table represent cubic feet of gas. Study the table carefully and you will realize that everything you need to know to use it correctly is right before your eyes.

Gas Pipe Sizing—IFG Table 402.4(2)
(Figure 4)

0.5 psi
0.5" WC
0.60 specific gravity

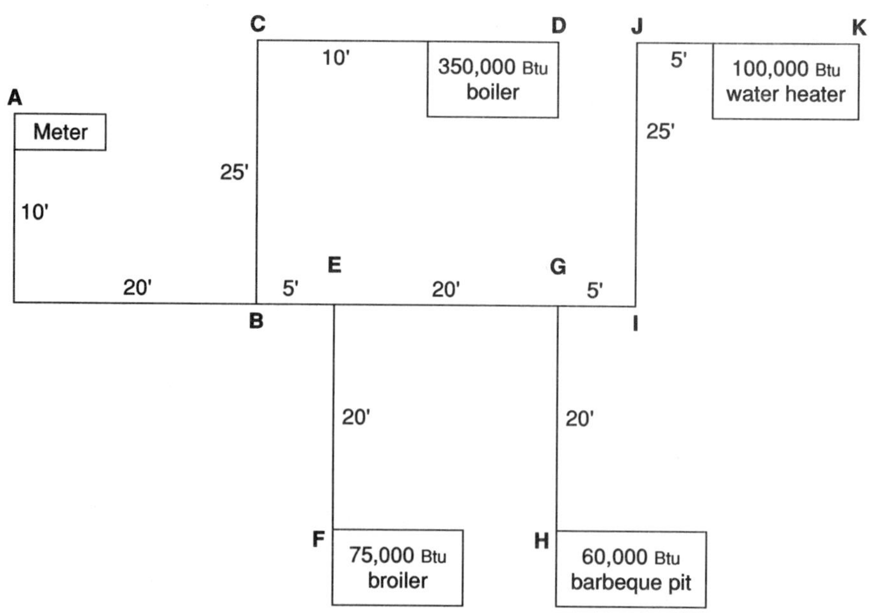

Questions 1–10 refer to Figure 4.

1. What is the minimum size gas pipe allowed to serve section A–B?

   a. ¾"

   b. 1"

   c. 1¼"

   d. 1½"

2. What is the minimum size gas pipe allowed to serve section B–C?

   a. ¾"

   b. 1"

   c. 1¼"

   d. 1½"

3. What is the minimum size gas pipe allowed to serve section C–D?

   a. ¾"

   b. 1"

   c. 1¼"

   d. 1½"

4. What is the minimum size gas pipe allowed to serve section B–E?

   a. ½"

   b. ¾"

   c. 1"

   d. 1¼"

5. What is the minimum size gas pipe allowed to serve section E–F?

   a. ⅜"

   b. ½"

   c. ¾"

   d. 1"

Notes

Notes

6. What is the minimum size gas pipe allowed to serve section E–G?

   a. ⅜"

   b. ½"

   c. ¾"

   d. 1"

7. What is the minimum size gas pipe allowed to serve section G–H?

   a. ⅜"

   b. ½"

   c. ¾"

   d. 1"

8. What is the minimum size gas pipe allowed to serve section G–I?

   a. ½"

   b. ¾"

   c. 1"

   d. 1¼"

9. What is the minimum size gas pipe allowed to serve section I–J?

   a. ½"

   b. ¾"

   c. 1"

   d. 1¼"

10. What is the minimum size gas pipe allowed to serve section J–K?

    a. ½"

    b. ¾"

    c. 1"

    d. 1¼"

PART SEVEN
General Knowledge

A portion of your exam will be based on general knowledge. These questions will be based on plumbing theory and the answers will be found in an approved textbook rather than the codebook. In identifying questions that are based on general knowledge, remember the codebook tells you what you must do and does not indicate how something is done. If the question includes the words *minimum*, *maximum*, *must be*, *shall be*, *etc.*, it is probably a code question. If the question is based on how and why something is done, the answer will be found in the approved reference book that discusses plumbing technology and theory.

Notes

GENERAL KNOWLEDGE PRACTICE EXAM #1

1. What is/are the required fixture/s for a family dwelling?

   a. Water closet
   b. Lavatory
   c. Tub/shower and sink
   d. All of the above

2. Plumbing systems in buildings include each of the following EXCEPT:

   a. Potable water
   b. Sanitary drainage
   c. Fuel gas piping
   d. Storm water

3. When planning a plumbing system, which of the following should be considered with regard to the piping?

   a. Available materials
   b. Local plumbing codes
   c. Cost of material, ease of installation, weight and durability of material
   d. All of the above

4. Name the four types of copper tube.

   a. Types schedule 80; schedule 40; schedule 30; schedule 10
   b. Types K; L; M; DWV
   c. Types heavy duty; service weight; with coupling; without coupling
   d. Types underground; above ground; in-wall; thin wall

5. What is galvanic action?
   a. Galvanic action is an electrolytic chemical reaction between dissimilar metals such as copper and iron
   b. Volcano action from a mountain or hill
   c. When sewer water enters a potable water system
   d. None of the above

6. What are the most common scales used in building drawings?
   a. ¼" = 1' and ½" = 1'
   b. ⅛" = 1" and ¼" = 1'
   c. ⅛" = 1' and ¼" = 1'
   d. ⅛" = 1' and ⅜" = 1'
   e. All of the above
   f. None of the above

7. What does this mean: SCALE ¼" = 1'0"
   a. Each ¼" of measurement equals 1 foot
   b. Each ¼" of measurement equals 1 inch
   c. Each 1 inch equals 4 feet
   d. Both A and C

8. Wall-hung water closet bowls shall be ________________.
   a. bolted securely to horizontal blocking
   b. connected by the supplied bolts to the wall
   c. rigidly supported by concealed metal supports
   d. None of the above

Notes

Notes

9. Plumbing codes are area laws established to control installation, maintenance, removal, and safety of plumbing conditions for the public and the plumbers or other craftsmen in the ________.

   a. city
   b. county
   c. state
   d. All of the above

10. Who issues licenses?

    a. Plumbing Inspector
    b. State and/or Local Authorities
    c. I.R.S.
    d. City Clerk

11. Each of the following is a requirement for plumbing fixtures EXCEPT:

    a. They shall be smooth surfaced
    b. They shall be made of non-absorbent material
    c. They shall be made of unbreakable material
    d. They shall be free of defects

12. What type of plugs do underground wyes and test tees use?

    a. Cast iron plugs
    b. Brass plugs
    c. Inflatable rubber test plugs
    d. Screw plugs

13. What does a testing gauge assembly do?
    a. It shows the pressure value to guarantee that your system is maintaining the rating required by that particular test
    b. It shows the required feet needed to give sufficient pressure
    c. It shows the slope needed to properly drain the system
    d. It shows the system being checked will pass the plumbing codes

14. What do the kitchen sink, garbage disposal, and dishwasher share?
    a. Single trap and drainpipe
    b. Water line
    c. 1¼" drain
    d. All of the above

15. What is a license?
    a. Permission for the plumber to begin working alone
    b. Displays their knowledge, experience and skill levels
    c. Both A and B
    d. You do not need a license to do plumbing

16. All water piping conveying nonpotable water shall be adequately and durably identified by a distinct _____ colored paint.
    a. red
    b. blue
    c. yellow
    d. None of the above

Notes

Notes

17. What is a permit?

    a. A permit grants permission for an individual to perform plumbing within the jurisdiction of the authority granting the permit

    b. A permit is an actual written document that grants permission for an individual or company to perform plumbing work at a predetermined site or building

    c. A permit grants permission for an individual or company to perform repair work on faucets

    d. A permit is an actual written document that lists all licensed plumbers in your area

18. What are the blueprint plans called that provide information about the plumbing, heating and air conditioning systems of a building?

    a. Plumbing

    b. Sepia

    c. Mechanical

    d. Architectural

19. What are some ways contamination can occur?

    a. Gravity backflow and cross-connections

    b. Back siphonage and backpressure

    c. Both A and B

    d. None of the above

20. On a gas water heater, what is a thermocouple?

    a. A small electric generator

    b. Fuel supply for pilot

    c. Water sensor

    d. Regulates tempature

21. A plumbing fitting which makes a 60 degree bend would also be known as a(n):

a. ⅛ bend

b. ⅙ bend

c. ⅕ bend

d. ¼ bend

22. What do you do if water hammering occurs in your water heater?

a. Install a water hammer arrester

b. Install an expansion tank

c. Install an air chamber

d. Either A, B or C

23. The use of primer and cement to make a joint in PVC pipe is known as:

a. Solvent welding

b. Insert fittings with hose and clamp

c. Air welding

d. Insert fittings with clamp rings

24. What is the difference between a sink and a lavatory?

a. Sinks are used as washing fixtures and a lavatory is a functional unit.

b. Sinks are functional units and a lavatory is used as a washing fixture

c. Sinks have three compartments and a lavatory has one compartment.

d. None of the above

Notes

Notes

25. What does ECO stand for when related to a water heater?

    a. Emergency control orifice

    b. Energy cutoff

    c. Both A and B

    d. None of the above

GENERAL KNOWLEDGE PRACTICE EXAM #2

Notes

1. What is an inspection?

   a. Inspections guarantee that the work fulfills code requirement and that the work is professional. An inspector will issue an "accepted" or "rejected" certificate

   b. Inspections give individuals permission to perform plumbing within the jurisdiction of authority granting the inspection

   c. Inspections make sure plumbers are doing quality work and that it is done according to code

   d. Both A and C

2. Which plumbing systems require testing?

   a. Potable water supply, distribution systems, and water services

   b. Storm drainage systems outside buildings

   c. Sanitary and venting systems, storm drainage systems inside buildings

   d. All of the above

   e. Both A and C

3. How are drainage and venting in buildings larger than 5 stories high tested?

   a. One test for all stories is O.K.

   b. Test each 5 story section separately with a 10' head of water.

   c. Test each story with a 10' head of water.

   d. Test all stories with 10 psi.

4. What prevents a chemical sink from being damaged?

   a. Continually running water

   b. Manufactured out of chemical-resistant materials

   c. P-trap seal

   d. Both B and C

   e. Both A and B

Notes

5. What do you have to do to a new garbage disposer before a dishwasher can be connected to it?

   a. Install a studor vent
   b. Install an air gap
   c. Knock out the drain plug located in the disposer drain outlet
   d. Nothing

6. What is the purpose of an anode rod on a water heater?

   a. The magnesium rod extends the tank life
   b. Makes the water hotter
   c. Controls the temperature
   d. Relieves pressure for safety

7. What is the purpose of a dip tube on a water heater?

   a. Transfers water from inlet to bottom of tank
   b. Filters out minerals from water
   c. Helps prevent backsiphonage by having a small hole at the top of the dip tube
   d. Both A and C

8. The desired air test pressure required for testing water supply lines:

   a. 1½ times the working pressure or 175 psi, whichever is the greater
   b. 2 times the working pressure or 175 psi, whichever is the larger
   c. Working pressure or 100 psi, whichever is the greater
   d. 1½ times the working pressure or 150 psi, whichever is the larger

9. The type of heating element used in electric water heaters:

   a. Low voltage elements
   b. 24 volt heating elements
   c. Resistance heating elements
   d. All of the above

10. A 160 foot sewer pipe installed with 1% grade will fall ____ inches.

    a. 60
    b. 19
    c. 28¾
    d. 30⅝

11. What should you do before connecting electrical wires or turning on the electrical breaker to an electric water heater?

    a. Check to see if connections are watertight
    b. Make sure all air is out of the tank by opening a hot water faucet and bleeding the air
    c. Make sure the water heater is level
    d. All of the above

12. What is a person called who has completed apprenticeship training in a particular field?

    a. Master
    b. Contractor
    c. Journeyman
    d. Worker

13. The primary purpose of a vent line in a water closet is ________.

    a. To remove odor from the building
    b. To prevent loss from trap seal
    c. To mix air with the sewage to help with the treatment
    d. To transport black water

Notes

Notes

14. Why is a relief valve so important to a water heater?

    a. It plugs a hole in the side/top of water heater

    b. It serves as a drain for the water heater

    c. It prevents super-heated water from exploding

    d. Both A and B

15. What are the two primary water sources for public potable drinking water?

    a. Potable water and contaminate water

    b. Surface water and ground water

    c. Rain water and well water

    d. All of the above

16. What is a cross-connection?

    a. When two separate piping systems one potable and one undesirable flow into each other

    b. When two dissimilar metals touch each other

    c. When two pipes cross

    d. None of the above

17. What is the hydrologic cycle?

    a. The continuous movement of water from the ocean, through the atmosphere, onto the land and back to the ocean

    b. The cycle completed when water is treated at the water works plant

    c. The chemical action when the rain enters the clouds

    d. Both A and B

18. Surface water enters the _________ through precipitation and infiltration.

    a. Streams
    b. Shallow wells
    c. Lakes and reservoirs
    d. All of the above

19. What is ground water?

    a. Water that runs off the ground
    b. Water that is brought to surface with wells and pumps
    c. Water located deep within the earth
    d. Both B and C

20. What is the chemical agent that softens and cleans the surface of plastic pipe and allows the cement to penetrate more effectively?

    a. Cement
    b. Primer
    c. Glue
    d. Pipe compound

21. Pipes passing through the foundation walls should be _______.

    a. Caulked
    b. Damp proofed
    c. Sleeved with cast iron pipe
    d. None of the above

Notes

Notes

22. What is used in soldering and brazing to facilitate the removal of oxides or undesirable substances from copper pipe?

    a. Flux

    b. Sandpaper

    c. Emery cloth

    d. Acid

23. In some areas, faucets installed after 1990 are required to be:

    a. Thermostatic or pressure-balanced

    b. Cartridge

    c. Single handle

    d. All of the above

24. When calculating single-pipe offsets, what is the vertical distance between the centers of the two offset fittings called?

    a. Travel

    b. Offset

    c. Advance

    d. None of the above

25. What device can be used to add a small amount of water to a floor drain trap to ensure an adequate trap seal?

    a. Ball valve

    b. Trap primer

    c. Trap valve

    d. Check valve

GENERAL KNOWLEDGE PRACTICE EXAM #3

Notes

1. The grade of a drainage pipe is ____________.

   a. The weight of the material
   b. The slope of the pipe
   c. The material sketch
   d. The abrasion resistance of the pipe

2. A temperature of at least ________ is necessary to braze joints and add the filler.

   a. 840°F
   b. 1,120°F
   c. 640°F
   d. 460°F

3. The inside diameter of a ¾ inches type L copper plumbing tube is:

   a. 0.811"
   b. 0.745"
   c. 0.785"
   d. 0.875"

4. What kind of pipe is joined together using a stainless steel clamp assembly surrounding a neoprene sleeve that seals the joint?

   a. PVC
   b. No-hub cast iron
   c. Hubbed cast iron
   d. None of the above

5. Which of the following is necessary for the purpose of directing water to the showerhead at tub and shower faucets?

   a. Cartridge
   b. Diverter
   c. Faucet washer
   d. Tub spout

Notes

6. What agency requires all employers to provide a safe environment for their employees?

   a. Environment Safety Association
   b. OSHA
   c. State Health & Safety Commission
   d. U.S. Board of Safety

7. A three-dimensional drawing of a plumbing system is referred to as:

   a. Schematic
   b. Floor plans
   c. Isometric
   d. Plan views

8. The most common individual vent is called:

   a. Straight
   b. Common
   c. Continuous
   d. Single

9. A vertical pipe installed to provide circulation of air to and from the drainage system is called:

   a. Vent stack
   b. Stack vent
   c. Main vent
   d. Main stack

10. What type of faucet uses a rubber washer pressing down on a brass seat to control the flow of water?

    a. Reverse-compression
    b. Compression
    c. Balancing control
    d. Cartridge

11. What is the drain fixture unit rating for a kitchen sink?

    a. 2 wsfu
    b. 2 dfu
    c. 1 dfu
    d. 3 wsfu

12. The fitting that mounts under the drain plug on a bathtub waste and overflow is called:

    a. Drain fitting
    b. Waste boot
    c. Tub elbow
    d. Waste shoe

13. The Energy Policy Act of 1992 mandated that the maximum volume of water used to flush a water closet could not exceed how many gallons?

    a. 2.1 gal.
    b. 3.5 gal.
    c. 1.6 gal.
    d. 6.1 gal.

Notes

Notes

14. What is a valve called that has a flush device actuated by direct water pressure to supply a fixed quantity of water for flushing purposes?

    a. Flushometer

    b. Ballcock

    c. Vacuum breaker

    d. Flushvalve

15. What is the drainage pipe extending horizontally from a soil or waste stack or building drain, with or without vertical sections or branches?

    a. Main drain

    b. Horizontal branch drain

    c. Building sewer

    d. None of the above

16. Approximately how much does one cubic foot of water weigh?

    a. 7.5 lbs.

    b. 26½ lbs.

    c. 36 lbs.

    d. 62½ lbs.

17. What provides a liquid seal in drainage piping to prevent the escape of sewer gases into the living area?

    a. Trap

    b. Fixture unit

    c. Vent system

    d. None of the above

18. What is the most commonly used type of pipe and fittings today?

   a. Cast iron
   b. Copper
   c. PVC
   d. Steel

19. The distance between the crown weir and the top of the trap Is:

   a. Crown length
   b. Weir distance
   c. Trap seal
   d. None of the above

20. What is caused when pressure in the water distribution pipe becomes higher than the water supply pressure?

   a. Backpressure
   b. Backsiphonage
   c. Cross-connection
   d. Atmospheric pressure

21. What is an air gap?

   a. Area between two pipes
   b. Atmospheric air between land and sky
   c. Distance between overflow tube and top of tank
   d. Unobstructed vertical space between the water supply and the flood rim of the receptacle

22. Expressed in inches, what is the distance the drainage pipe drops in its given length?

   a. Slope
   b. Fall
   c. Grade
   d. Both A and C

Notes

Notes

23. What is the faucet installed on the outside of a building called?

    a. Check valve
    b. Sillcock
    c. Pressure vacuum breaker
    d. Water shutoff valve

24. What is the valve with hose threads installed to drain a water heater called?

    a. Boiler drain
    b. Hydrant
    c. Hose bibb
    d. Water shutoff

25. A valve that permits fluid flow in only one direction and closes automatically to prevent backflow is referred to as:

    a. Check valve
    b. Anti-siphon valve
    c. Ball valve
    d. All of the above

GENERAL KNOWLEDGE PRACTICE EXAM #4

Notes

1. What is a valve that converts high and/or fluctuating inlet water pressure to a lower or constant outlet pressure called?

   a. High pressure valve
   b. Backpressure valve
   c. Pressure-reducing valve
   d. Both A and C

2. What type of traps do water closets have?

   a. P-trap
   b. Integral
   c. Bell trap
   d. Crown weir trap

3. In a flush tank assembly, how high is the water level in comparison to the overflow tube?

   a. The level should be ¾" below top of overflow tube
   b. The level should be at the top edge of overflow tube
   c. The level should be 1½" below top of overflow tube
   d. The level makes no difference.

4. In a flush tank assembly, what is the purpose of the ballcock assembly?

   a. Acts as an anti-siphon valve
   b. Fills tank with water
   c. Lets in the water
   d. All of the above

Notes

5. What is the maximum flush volume for a urinal?

   a. 1.0 gpm
   b. 1.6 gpm
   c. ¾ gpm
   d. 3.5 gpm

6. PE pipe (polyethylene) is black and is typically sold:

   a. In 10' lengths
   b. By the foot
   c. In coils
   d. In 21' lengths

7. PE pipe (polyethylene) is used for:

   a. Jet wells and water supply systems
   b. Farm sprinkler systems
   c. Salt lines, chemical well lines and gas gathering systems
   d. All of the above

8. How is PE (polyethylene) joined together?

   a. Plastic insert fittings and clamps
   b. No-hub clamps
   c. Cement and primer
   d. Compression couplings

9. Typically, manufacturers guarantee PVC pipe for:

   a. 10 years
   b. 50 years
   c. 75 years
   d. 100 years

10. One pint of plastic pipe cement will fuse approximately how many 1½ inch joints?

    a. 35
    b. 25
    c. 100
    d. 50

11. What color cement is used for CPVC pipe and fittings?

    a. Clear
    b. Orange
    c. Black
    d. Blue

12. What is the color difference in PVC and ABS pipe and fittings?

    a. PVC is black and ABS is white
    b. PVC is white and ABS is black
    c. PVC is plastic and ABS is cast iron
    d. PVC is schedule 40 and ABS is schedule 30

13. Can both PVC and ABS pipe and fittings be joined together with PVC cement?

    a. No, they are different types of plastic pipe
    b. Yes, PVC cement will work for any type of plastic
    c. Yes, but only for these two types of plastic
    d. Yes, as long as you use the correct primer

14. Can PVC pipe and fittings be used for the hot water system?

    a. Yes, PVC can be used on hot or cold water systems
    b. No, use CPVC or pipe that is approved for hot water systems
    c. No, use only copper or galvanized steel pipe
    d. None of the above

Notes

Notes

15. What is the quietest drain pipe to install in a building?

    a. PVC plastic pipe
    b. Copper DWV pipe
    c. Steel pipe
    d. Cast Iron pipe

16. The drawing scale of a drawing is ⅛" = 1'0". The length of the pipe measures 2¼" on the blueprint. How long is the actual length of the pipe?

    a. 16 ft.
    b. 18 ft.
    c. 18 in.
    d. 20 in.

17. On blueprints, what is the purpose of phantom lines?

    a. Used to display water lines
    b. Used to display buried pipes
    c. Used to display drainage and vent lines
    d. Used to display moveable parts

18. On a blueprint, what does a "key" tell you?

    a. Keys describe what the lines or specific abbreviations represent
    b. Keys tell where the fixtures are located
    c. Keys show where to install drainage pipes
    d. All of the above

19. The inside diameter of a 1½", schedule 40, nominal pipe size is:

    a. 1.90"
    b. 0.145"
    c. 1.610"
    d. 0.200"

20. A three-inch short sleeve turn may be used in all but which of the following locations?

    a. In waste lines with a change in flow in horizontal to vertical
    b. In waste lines with a change in flow in vertical to horizontal
    c. In waste lines between floors
    d. In waste lines where the change in flow is from horizontal to horizontal

21. Ejector pumps may be connected to a ____________.

    a. Wye fitting in the building drain
    b. Cleanout at the upstream end of the building sewer
    c. Sanitary sewer and the building drain
    d. Tee fitting and a branch of the building drain

22. Single sanitary tees, double sanitary tees, and quarter bends, may be used in vertical sections of drainage lines only where the direction of the flow is ____________.

    a. From horizontal to vertical
    b. From vertical to horizontal
    c. Less than 20 degrees off horizontal
    d. Less than 45 degrees off horizontal

23. Drops for lavatories are made up to be ___ inches long.

    a. 24
    b. 30
    c. 36
    d. 42

Notes

Notes

24. The blowout water closet is a commercial type water closet that depends on a large jet of water directed into the _____ of the trap.

    a. inlet

    b. outlet

    c. weir

    d. None of the above

25. When threading 1" steel pipe, the recommended number of threads per inch is:

    a. 11

    b. 11.25

    c. 11.5

    d. 11.75

GENERAL KNOWLEDGE PRACTICE EXAM #5

Notes

1. Which are typically used to fasten or secure pipe supports to structural components of wood frame construction?

   a. Plastic inserts

   b. Expansion plugs

   c. Nuts and bolts

   d. Nails and screws

2. Trap siphonage and back siphonage pressure are a result of pressure differences between the drainage system and ___________.

   a. ventilation duct

   b. extension trap

   c. biosphere

   d. atmosphere

3. Branch lines which enter a stack at or near each floor level in a building is known as a ______.

   a. vent stack

   b. waste stack

   c. branch interval

   d. branch soil pipe

4. Which of the following can only exist where there is another floor above the one in question?

   a. Wet vent

   b. Loop vent

   c. Circuit vent

   d. Yoke vent

Notes

5. Which of the following are plumbing pipes made from?

   a. Plastic

   b. Copper

   c. Cast iron soil pipe

   d. All of the above

6. Water distribution pipes are generally represented by which color?

   a. Green

   b. Yellow

   c. Tan

   d. None on the above

7. A vent pipe that connects at the junction of two fixtures drains and serves as a vent is known as a(n)______.

   a. relief vent

   b. back vent

   c. common vent

   d. individual vent

8. A vent stack is connected to the waste stack at the lower end with a wye and _______.

   a. 1⁄16 bend

   b. 1⁄8 bend

   c. 1⁄6 bend

   d. 1⁄4 bend

9. In a multi-story building, the waste stack and vent stack must reconnect with a yoke vent every ______ branch intervals.

   a. 4

   b. 5

   c. 6

   d. 8

10. Strainer baskets attached to a building storm drain to prevent debris from entering storm water systems are typically made of ______.

    a. cast iron

    b. cast aluminum

    c. plastic

    d. All of the above

11. A trap primer is a device which adds a small amount of water to the floor drain trap to ensure an adequate trap seal. Which of the following is not a type of trap primer?

    a. Dispensing trap primer

    b. Automatic trap primer

    c. Water saver trap primer

    d. Vacuum breaker trap primer

12. In a multi-family dwelling, common fixtures are installed ______.

    a. back to back

    b. side to side

    c. front to back

    d. front to front

13. Which of the following is a prohibited trap if the inlet and outlet of the trap are in vertical alignment with each other?

    a. Gag

    b. Bag

    c. Slug

    d. Key

Notes

Notes

14. What factor can be affected by a plumber when installing a plumbing system?

    a. Number of fixtures

    b. Size of piping

    c. Type of fittings

    d. Length of piping

15. Which of the following receives only liquid body waste and conveys the waste through a trap seal into a sanitary drainage system?

    a. Water closet

    b. Bidet

    c. Lavatory

    d. Urinal

16. How many bolts are required to anchor a siphon jet water closet bowl to the supporting chair carrier?

    a. 2

    b. 3

    c. 4

    d. 5

17. A floor set basin used for maintenance tasks and cleaning is known as a______.

    a. floor sink

    b. utility sink

    c. maintenance basin

    d. mop basin

18. Which of the following physical properties affect water in supply lines?

    a. Density
    b. Boiling point
    c. Viscosity
    d. All of the above

19. A lawn sprinkler is connected to the potable water service supply pipe, there must be an approved backflow preventer on the discharge side of each valve. The backflow preventer must be at least ________.

    a. 6 inches above the highest sprinkler head
    b. 6 inches away from the highest sprinkler head
    c. 8 inches above the highest sprinkler head
    d. 8 inches away from the highest sprinkler head

20. Which of the following must include a yard hydrant because of their size, isolation, or type of occupancy?

    a. Mobile home park
    b. Marinas
    c. Lumber yards
    d. All of the above

21. Pedestal urinals must have a minimum of ______ inches center to center spacing when set in a battery installation.

    a. 15
    b. 18
    c. 30
    d. None of the above

Notes

Notes

22. Plastic piping from a septic tank to a drainfield located under paving shall be?

    a. ABS schedule 40
    b. PVC schedule 40
    c. PVC schedule 80
    d. CPVC piping only

23. On a loop system a vent shall be installed vertically downstream, and be within ______ feet of the first fixture branch.

    a. 5
    b. 7
    c. 8
    d. 10

24. Which of the following materials shall not be used for storm drainage within a building?

    a. Bituminized fiber pipe
    b. Galvanized steel pipe
    c. Plastic pipe
    d. Lead pipe

25. Which of the following hose connected faucets are not required to have a backflow preventer installed?

    a. Three compartment commercial sink
    b. Service sink
    c. Automatic clothes washing machine
    d. Outside hose faucet

GENERAL KNOWLEDGE PRACTICE EXAM #6

Notes

1. The process of joining two pieces of metal using heat and the application of another metal of different composition is called ______.

   a. brazing
   b. soldering
   c. compression
   d. face feeding

2. Capped pipe nipples that penetrates walls and floors to help locate supply and drain lines are known as ______.

   a. cleanouts
   b. location stubs
   c. stub pieces
   d. install fixtures

3. What type of faucet is sometimes used in public buildings specifically to conserve water?

   a. Self closing faucet
   b. Compression faucet
   c. Quick pression faucet
   d. Quick opening faucet

4. Which of the following slopes is the most common for main house drains?

   a. 1⁄16" per foot
   b. 1⁄8" per foot
   c. 1⁄4" per foot
   d. 1⁄2" per foot

Notes

5. A magnesium rod used to retard corrosion in a water heater is called a ______.

   a. dip tube

   b. stabilizer rod

   c. adapter tube

   d. sacraficial rod

6. A relief valve is used to prevent damage from excessive pressure in a closed water, air, or steam system. A small relief valve, ½" to 2", is made of ______.

   a. brass with threaded connections

   b. cast iron

   c. steel with brass seats

   d. All of the above

7. Which of the following is true about a water-filled trap?

   a. Siphonage will upset trap seal

   b. Back pressure will upset trap seal

   c. Both a and b

   d. Neither a or b

8. What prevents waste from slowing down at the base of a stack and forming a slug of water that would cause pressure changes?

   a. Aerator

   b. Deaerator

   c. Separation chamber

   d. None of the above

Notes

9. Which type of drain fitting is commonly installed on bathtubs?

   a. Waste fitting
   b. Overflow fitting
   c. Diverter fitting
   d. Lift waste fitting

10. Which of the following fixtures is not considered when if installed with 25' of vented drainage pipeline?

   a. Lavatory
   b. Bidet
   c. Water closet
   d. Floor drain

11. Which of the following is not true when considering minimum requirements of a floor drain?

   a. Floor drain strainer must be removable
   b. Minimum outlet size of a floor drain is 1½"
   c. The combined free area of the holes in the grate must equal the drain outlet size
   d. Must be equipped with a backwater valve when installed below ground level

12. All of the following are necessary when installing a drinking fountain that discharges indirectly into the drainage system except________.

   a. 1¼" vent pipe
   b. 1¼" P-trap
   c. 1¼" waste pipe
   d. All of the above

Notes

13. Scaled views that show the shape and size of the building lot and the location, shape, and overall size of a house is known as _______.

    a. plot plans
    b. details
    c. sectional views
    d. floor plans

14. Which of the following results from negative pressure in the distribution pipes of a potable water supply system?

    a. Back pressure
    b. Back siphonage
    c. Flooding
    d. None of the above

15. Convert the following into feet: 10 feet 10 inches

    a. 10.10'
    b. 10.12'
    c. 10.83'
    d. 130"

16. What is the capacity, in gallons, of a 2 inch pipe that measures 10 feet long? Given: Volume $= 0.0408 \times D^2 \times L$

    a. 0.82 gal.
    b. 1 gal.
    c. 1.25 gal.
    d. 1.63 gal.

17. Horizontal drainage pipeline is designed to flow about ______ full.

a. ⅓
b. ⅔
c. ¾
d. ½

18. Which type of fitting has the same dimensions at all openings?

a. Straight
b. Standard
c. Reducing
d. None of the above

19. Which of the following is not true about gas piping?

a. Aluminum pipe will not be used outside or underground
b. Plastic pipe may only be used inside
c. Aluminum pipe must be factory coated where it may be in contact with detergent
d. All of the above

20. Septic tanks should be pumped out every ______ years.

a. 1 to 2
b. 3 to 5
c. 5 to 7
d. 8 to 10

21. All porcelain enamel surfaces on plumbing fixtures shall be ______.

a. shatter resistant
b. acid resistant
c. heat resistant
d. freeze resistant

Notes

Notes

22. Which of the following is considered to be hazardous material?

    a. Thread cutting oil
    b. Cleaning solvent
    c. Solvent cement
    d. All of the above

23. Decomposition of organic material found in a sewer is a hazardous by-product called ______.

    a. sludge
    b. soil
    c. sewer gas
    d. oxygen

24. A(n) ______ plug is a test device installed on the inside of the pipe end.

    a. inspection
    b. pipe
    c. gauge
    d. inlet

25. A(n) ______ cap is a test device installed on the outside of a pipe end.

    a. pipe
    b. gauge
    c. tee
    d. outside

"IDENTIFY" QUIZ

1. Identify this tool.

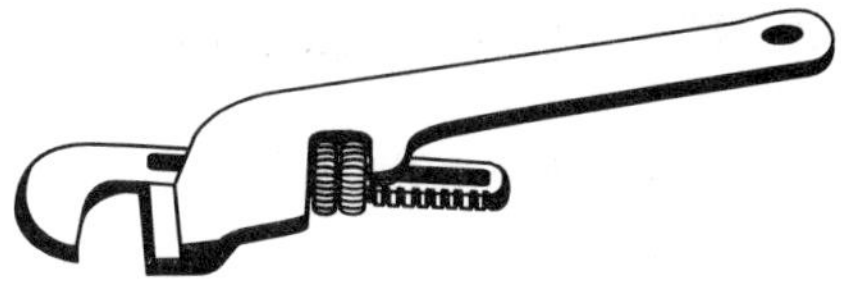

   a. End pipe wrench
   b. Straight pipe wrench
   c. Monkey wrench
   d. Adjustable wrench

2. Identify the valve.

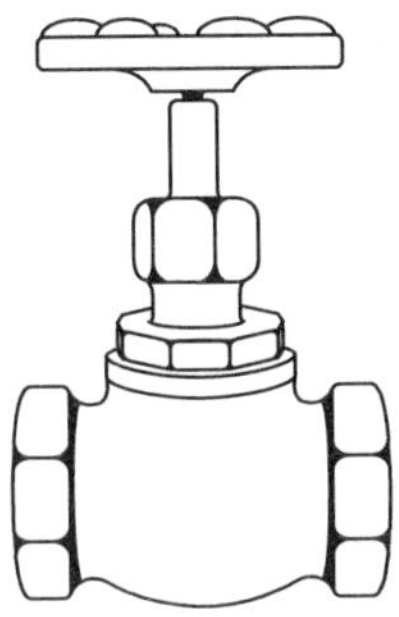

   a. Gate
   b. Globe
   c. Angle
   d. Rising stem

3. Identify this tool.

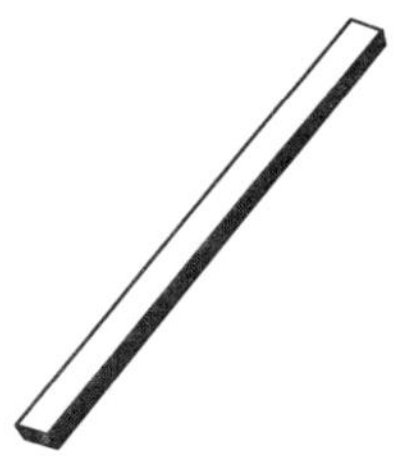

   a. Graphite placement device
   b. Pen
   c. Pencil
   d. Soap stone

4. Identify this fixture.

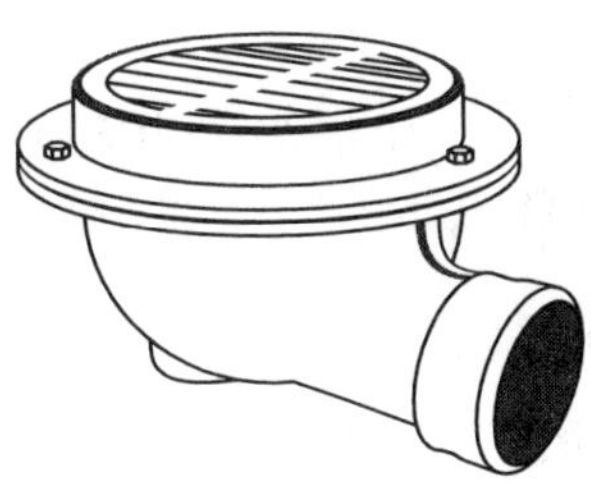

   a. Bottom outlet floor drain
   b. Bottom outlet roof drain
   c. Side outlet floor drain
   d. Side outlet roof drain

5. Identify this tool.

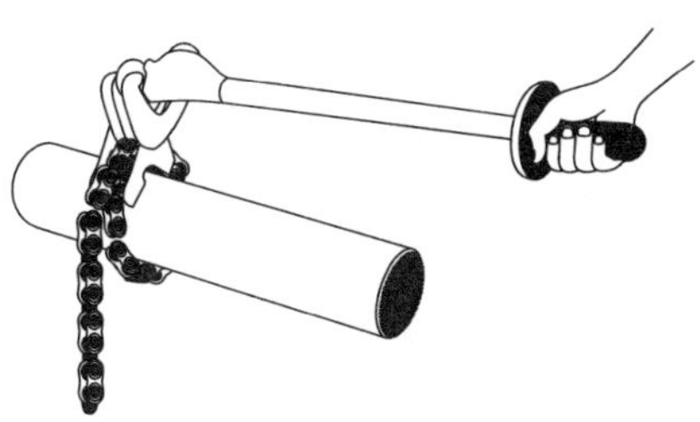

   a. Squeeze type cutter
   b. Chain ratchet cutter
   c. Hydraulic cutter
   d. Hand cutter

Notes

Notes

6. Identify this tool.

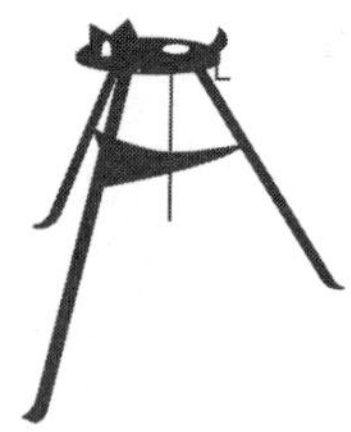

   a. Pipe reamer
   b. Pipe cutter
   c. Pipe threader
   d. Pipe stand

7. Identify the fitting.

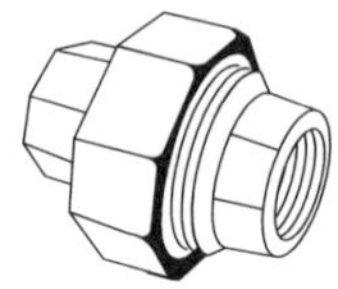

   a. Male adaptor
   b. Plug
   c. Reducer
   d. Union

8. Identify this fitting.

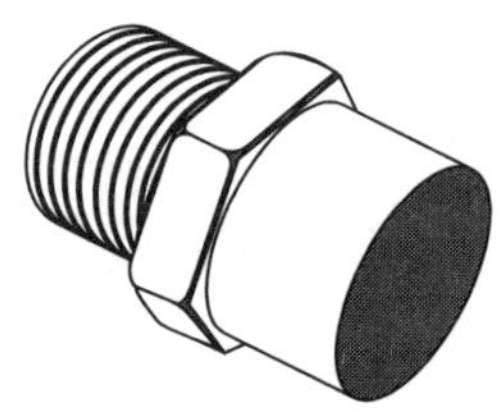

   a. Copper to FIP adapter
   b. Copper coupling
   c. Copper union
   d. Copper to MIP adapter

9. Identify this fitting.

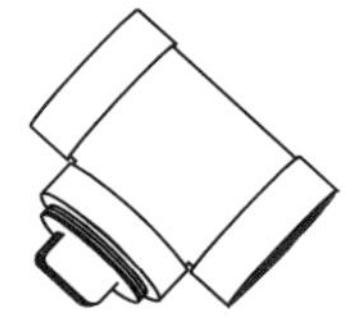

   a. Vent tee
   b. Sanitary tee
   c. Sweep tee
   d. Cleanout tee

10. Identify this tool.

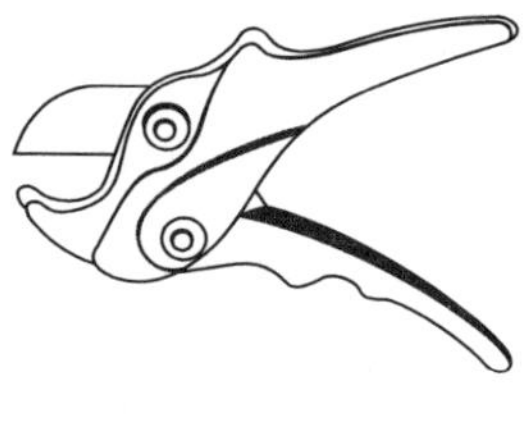

   a. Universal saw
   b. Ratchet cutter
   c. Tubing cutter
   d. Crimping tool

11. Identify the fitting.

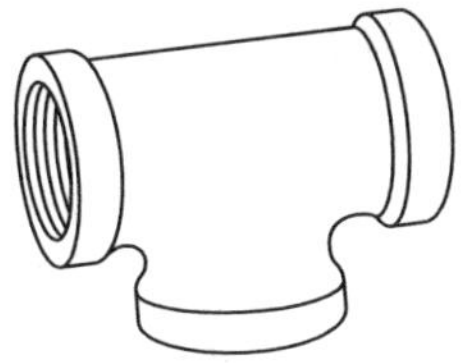

   a. Bell and spigot tee
   b. Solvent welded tee
   c. Screwed tee
   d. Insert tee

12. Identify this tool.

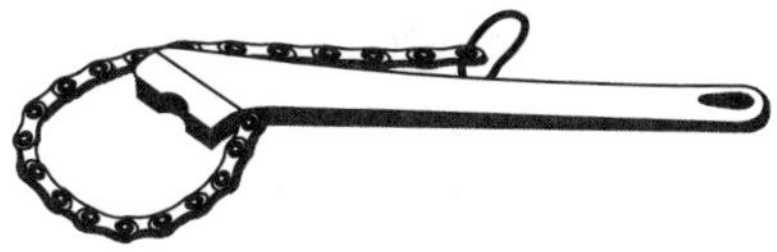

   a. End pipe wrench
   b. Strap pipe wrench
   c. Compound wrench
   d. Chain wrench

13. Identify the fitting.

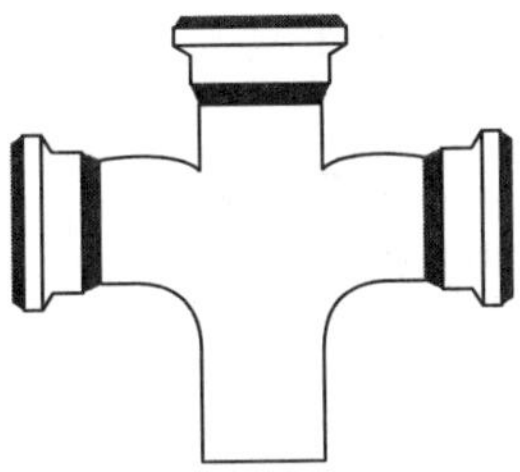

   a. Bell and spigot cross
   b. Double sanitary tee—no hub
   c. Plastic cross
   d. PVC cross

14. Identify this tool.

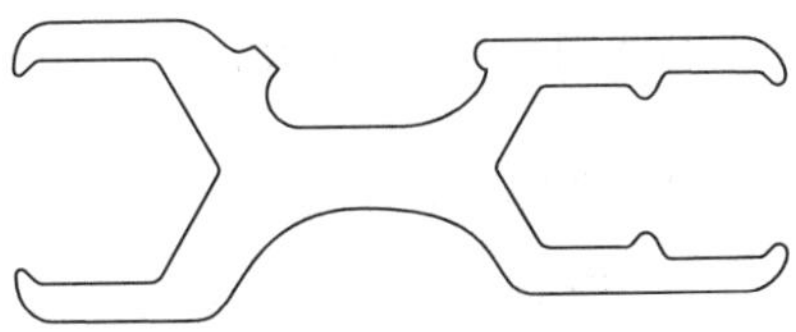

   a. Smooth-jawed pliers
   b. Spud wrench
   c. Offset hex wrench
   d. Adjustable wrench

15. Identify this tool.

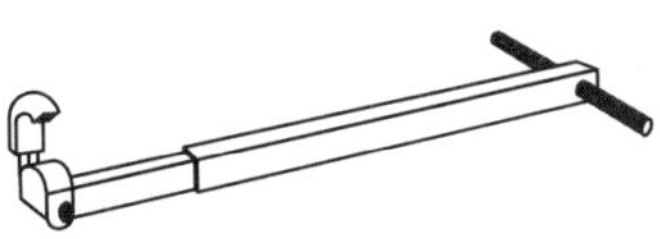

   a. Tub wrench
   b. Sink wrench
   c. Basket strainer wrench
   d. Basin wrench

Notes

Notes

16. Identify this tool.

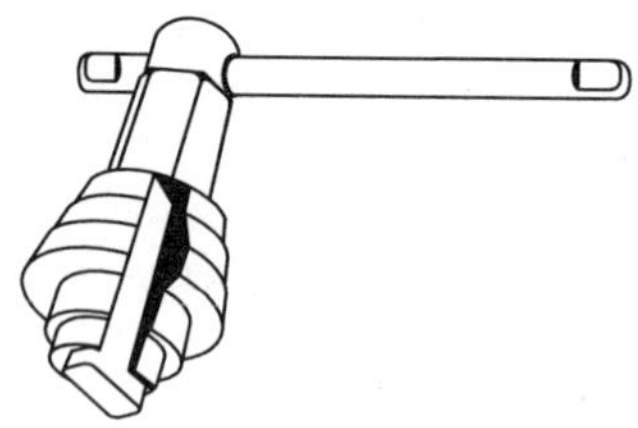

   a. Monkey wrench
   b. Tub wrench
   c. Internal wrench
   d. Offset hex wrench

17. Identify this tool.

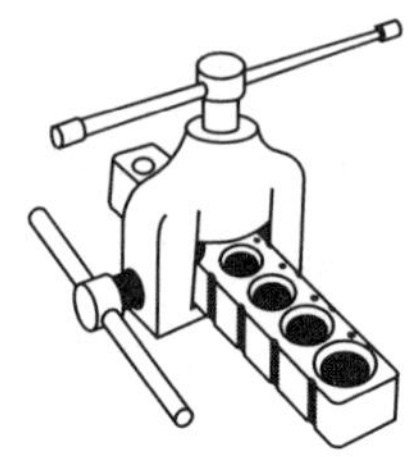

   a. Flaring tool
   b. Adjustable wrench
   c. Tube cutter
   d. Pipe reamer

18. Identify this tool.

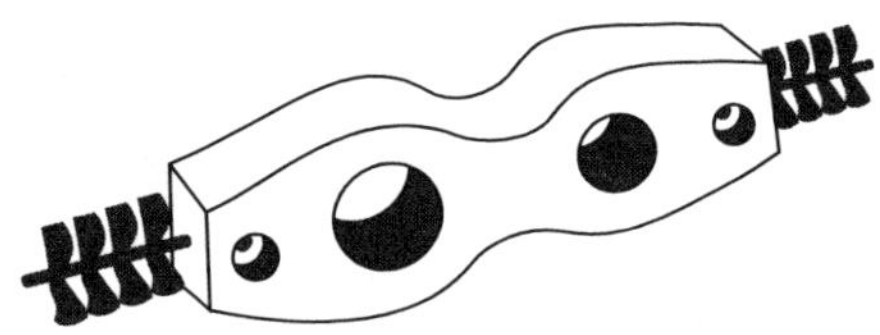

   a. Inner-outer copper cleaning tool
   b. Copper pipe reamer
   c. Copper tube cutter
   d. Copper fitting brush

19. Identify this tool.

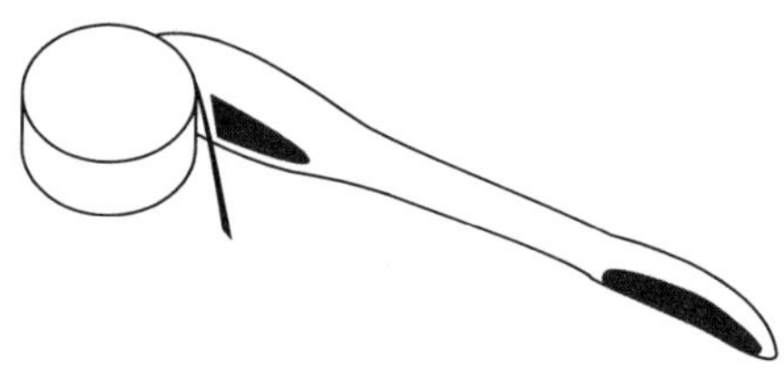

   a. Pipe wrench
   b. Crescent wrench
   c. Strap wrench
   d. Chain wrench

20. Identify this tool.

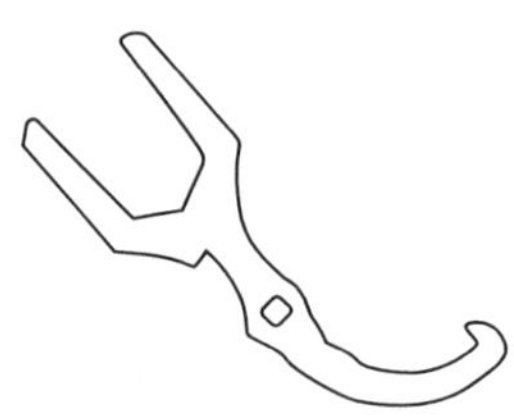

   a. Spud wrench
   b. Sink drain wrench
   c. Garbage disposal wrench
   d. Internal wrench

21. Identify this tool.

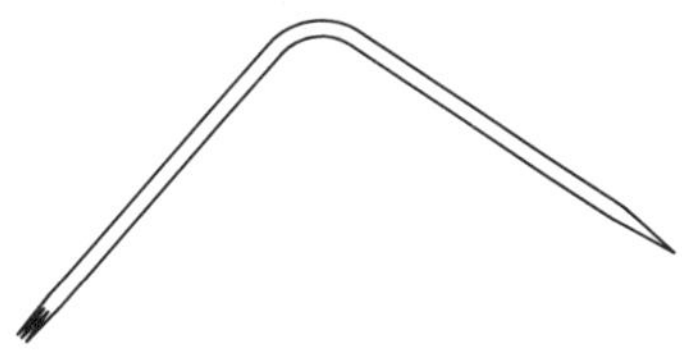

    a. Delta faucet repair wrench
    b. Garbage disposal wrench
    c. Allen wrench
    d. Faucet seat removal wrench

22. Identify this tool.

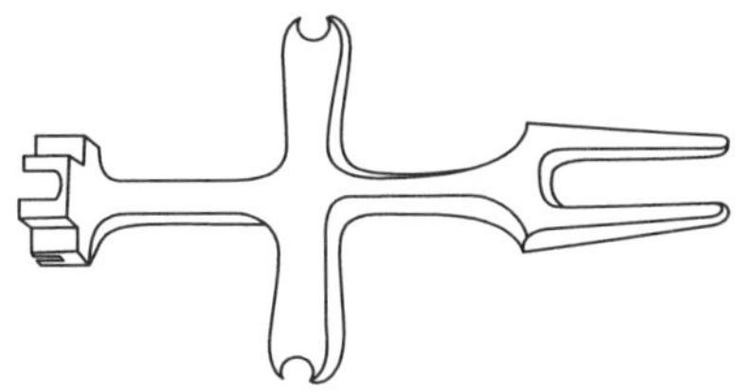

    a. Flushometer valve wrench
    b. Fit-all wrench
    c. PO plug wrench
    d. Check valve repair wrench

23. Identify this tool.

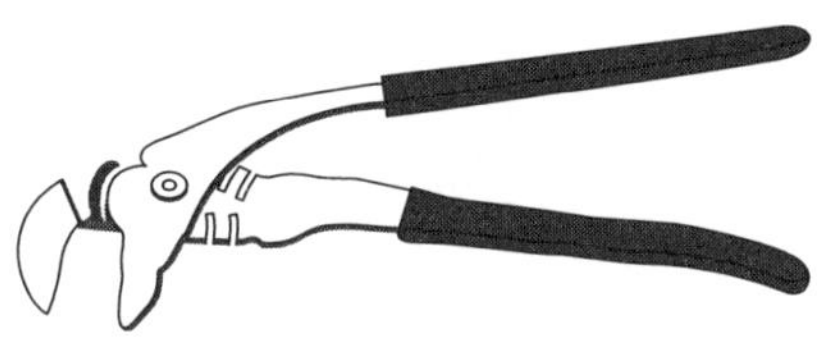

    a. Smooth-jaw pliers
    b. Monkey wrench
    c. Water pump pliers
    d. Slip-joint pliers

24. Identify this tool.

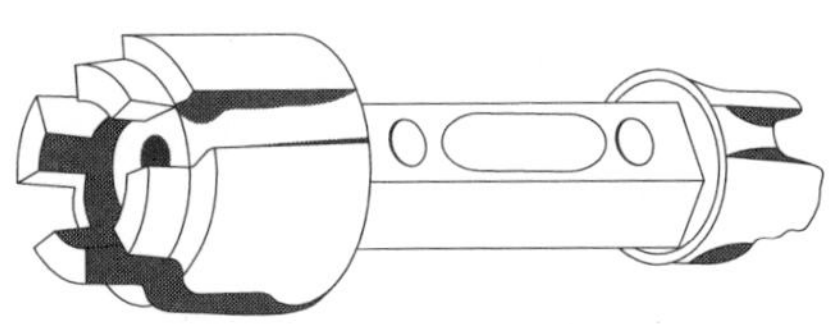

    a. Sink basket strainer wrench
    b. PO plug wrench
    c. Tub drain wrench
    d. Fit-all wrench

25. Identify this tool.

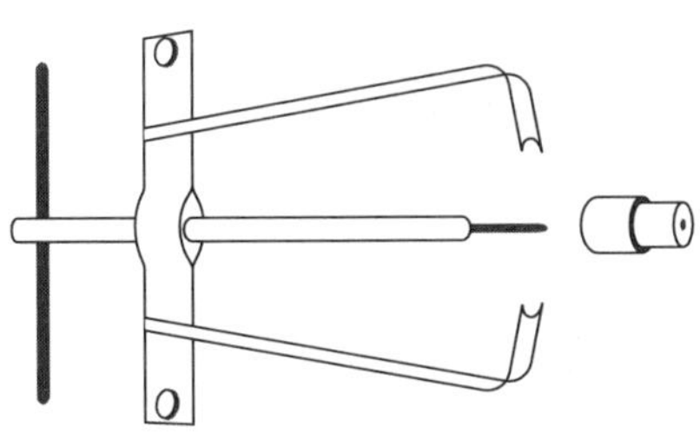

    a. Faucet handle puller
    b. Seal puller
    c. Moen cartridge puller
    d. None of the above

Notes

Notes

PART EIGHT
Project Management

The following questions are typical of those found in the Project Management section of most state contractor licensing exams. While many state requirements include separate exams to cover this content, several states incorporate these questions intermittently with the trade questions. Be careful to read your exam candidate bulletin to determine whether your exam covers this subject matter and to determine the approved reference material.

Notes

PROJECT MANAGEMENT PRACTICE EXAM #1

1. Each of the following is regulated by the Fair Labor Standards Act EXCEPT:

   a. Vacation

   b. Time and a half for hours over 40

   c. Minimum wage

   d. Compliance with child labor laws

2. What amount does an employer pay for Social Security and Medicare?

   a. 6.2%

   b. 1.45%

   c. 7.65%

   d. 5.4% on the first $7,000

3. Two brothers are planning to start a plumbing contracting company together. The oldest brother will be the financial backer primarily and wants no day-to-day involvement in the operations. What type of business should the brothers form?

   a. General partnership

   b. Proprietorship

   c. Limited partnership

   d. Corporation

4. The best source for projecting future project and company overhead percentage rates is:

   a. Historical data

   b. Bid documents

   c. Unit pricing

   d. Job cost records

5. Which of the following insurance types is most widely used for construction projects?

   a. Building operations
   b. Liability
   c. All-risk builders-risk
   d. Catastrophe

6. Which of the following would be considered project overhead?

   a. Yellow page advertising
   b. Secretary's salary
   c. Electric bill for the office
   d. Electric bill for project office

7. Yellow Duck Plumbing Company has 21 employees. According to OSHA, this company must keep a record for _______ years after each recorded accident for each one of its employees while on the job site.

   a. 3
   b. 4
   c. 5
   d. 6

8. What entity continues to exist beyond the death of one of the principals?

   a. Corporation
   b. Partnership
   c. Sole proprietorship
   d. Limited partnership

Notes

Notes

9. Employers are not allowed to deduct _______ from employee's wages.

   a. Social Security
   b. Medicare
   c. Federal Unemployment Tax (FUTA)
   d. Federal Income Tax

10. Which of the following is a disadvantage of setting up a general partnership?

    a. Single taxation
    b. Easy setup
    c. Unlimited personal liability
    d. The inability to pool financial resources

11. Which of the following would NOT constitute a material breach of contract?

    a. Building on the wrong site
    b. A refusal to work
    c. The owner refuses to pay
    d. A delay in the work

12. What form of business organization offers contractors the most protection against personal liability?

    a. Corporation
    b. Partnership
    c. Joint venture
    d. Sole proprietorship

13. Given: A carpenter drops a hammer from a roof and injures a curious bystander. Which type of insurance would cover the victim's medical costs?

    a. Builder's risk
    b. Named-peril risk
    c. Liability
    d. Risk management

14. The simplest form of business ownership is:

    a. Corporation
    b. Partnership
    c. Limited partnership
    d. Sole proprietorship

15. What form is used to report Federal Unemployment taxes (FUTA)?

    a. W-3
    b. 940
    c. W-2
    d. 1099

16. A large plumbing contractor maintains a central office and has crews working at several projects. Which of the following is considered a company overhead cost?

    a. Superintendent salary
    b. Temporary barricades
    c. Site cleanup
    d. Advertising

Notes

Notes

17. Typically, the only party with whom the subcontractors and their employees have a contractual relationship with is the:

    a. Owner's representative

    b. Architect

    c. Owner

    d. Contractor

18. The purpose of the journal:

    a. To post transactions on a weekly basis

    b. To post transactions on a daily basis

    c. To ensure proper job costing

    d. To balance the checkbook

19. What is the latest date an employer can file the Federal Unemployment Tax Return (Form 940)? (Assume no time extensions are taken and the balance due is greater than $100.00)

    a. December 1

    b. December 31

    c. January 1

    d. January 31

20. Employee records regarding pay shall be kept for at least _____ years.

    a. 1

    b. 2

    c. 3

    d. 4

21. Which of the following is the proper formula for determining ASSETS?

    a. Liabilities + Owner's equity

    b. Liabilities − Owner's equity

    c. Owner's equity − Liabilities

    d. Assets − Liabilities

22. Which of the following components make a change order complete?

    a. The change to be made, the charge, and the effect on time
    b. The owner's/contractor's address, and the change order number
    c. The legal description and the contractor's license number
    d. The contractor's license number, the change order number, and the architect's signature

23. The primary purpose of a written contract is to:

    a. Define rights and obligations of each party
    b. Standardize all requirements of work performance
    c. Condense construction requirements
    d. Standardize all requirements of work performance

24. The term used in scheduling that makes reference to the TIME a project can be delayed without extending the project completion:

    a. Float time
    b. Lag time
    c. Critical task
    d. Non-critical task

25. Which of the following terms is used to describe withholding a percentage of a progress payment pending completion to a certain point for the purpose of covering defects or faulty installation?

    a. Retainage
    b. Withholding
    c. Holdback
    d. None of the above

Notes

Notes

PROJECT MANAGEMENT PRACTICE EXAM #2

1. A bid bond provides ______________.

   a. Assurance that the bonding company is liable for all bid discrepancies

   b. Assurance to the owner or government agency that the contractor will honor the bid submitted by entering into a contract in accordance with the terms of the bid

   c. A guarantee to the owner that the project will be completed

   d. A guarantee of both job completion and payment of all labor and materials

2. A truck that is due to be paid off within a year would be considered a(n):

   a. Current liability

   b. Long-term liability

   c. Asset

   d. Other

3. In the event a subcontractor does not carry worker's compensation insurance on his employees, who is responsible for providing the insurance?

   a. Owner

   b. General contractor

   c. State government

   d. No one

4. Which of the following financial statements indicate the financial condition of a company at a given point in time?

   a. Income statement

   b. Cash flow statement

   c. Balance sheet

   d. Funds flow statement

5. A first aid kit is checked before it is sent to the job, and then checked _______________.

   a. Every week
   b. Bi-weekly
   c. Monthly
   d. Quarterly

6. Which of the following would be classified as a long term debt?

   a. Current annuity of long term debt
   b. Mortgage on company office building
   c. Accounts payable
   d. Accumulated depreciation

7. What is another name for owner's equity?

   a. Operating capital
   b. Liquid assets
   c. Net worth
   d. Liability

8. Which of the following statements defines *Networking* Capital?

   a. The amount the current assets exceeds the current liabilities
   b. The amount listed as an item overhead expense
   c. The difference between the direct cost and the cost of sales
   d. Current liabilities less overhead expense

9. If two companies enter into a project as a joint venture, if one company does not have a contractor's license, what must occur?

   a. Company with license will manage
   b. A license in the name of the joint venture shall be obtained
   c. This is permitted as long as at least one firm has a license
   d. Other

Notes

Notes

10. Based on an accrual basis of accounting; earnings, which have not been received, will be displayed as ____________.

    a. Current assets
    b. Fixed assets
    c. Current liabilities
    d. Owner's equities

11. The required quality of a project is described in the:

    a. Budget
    b. Specifications
    c. Plans
    d General conditions

12. Current assets are assets that can be converted into cash within ______.

    a. 1 year
    b. 6 months
    c. 1 month
    d. 2 years

13. A form of business organization that has a tax structure similar to a partnership but is organized as a corporation:

    a. Limited partnership
    b. Subchapter S corporation
    c. General partnership
    d. Non-profit organization

14. Regarding contracts, which of the following statements is NOT true?

    a. Words prevail over numbers

    b. General clauses prevail over special clauses

    c. When properly documented, handwritten agreements prevail over typed provisions.

    d. A contract deemed ambiguous is interpreted against the party that provided it.

15. Which of the following work related injuries does NOT require the employer to report to OSHA?

    a. A broken wrist

    b. An X-ray

    c. Cancer due as a result of long-term exposure to work-related chemicals

    d. A sprain that removed an employee from work for 2 days

16. What is a benefit of good inventory control?

    a. Better for safety

    b. Better preparation for increase in business

    c. Reduced time and expense in handling materials

    d. Ability to ship faster to offsite storage

17. Periodic payments made by the owner to the contractor over a long project are called:

    a. Progress payments

    b. Acceleration payments

    c. Performance payments

    d. Provisional payments

Notes

Notes

18. What accounting method records income when cash is received?

    a. Accrual basis
    b. Cash basis
    c. Single entry
    d. Double entry

19. What portion of the worker's compensation insurance premium can an employer charge to an employee?

    a. A maximum of 2% of the employee's weekly gross earnings
    b. No deduction is allowed
    c. Up to 50% of the premium
    d. A flat fee of $25 per week

20. If an employee is employed by a corporation and takes the licensing examination for the corporation, who will own the license when it is issued?

    a. The employee
    b. The stockholders
    c. The president of the corporation
    d. The corporation itself

21. An employer must maintain records about work related injuries for _______.

    a. 3 years
    b. 5 years
    c. 1 year
    d. 7 years

22. Which of the following statements is NOT true regarding a bar chart method of scheduling?

    a. Shows the start and finish times for each task

    b. Does not show how tasks relate to each other

    c. Most common scheduling system found on the job site

    d. None of the above (all are true)

23. When a clause in a contract calls for liquidation damages, it is referring to:

    a. A payment per day for finishing after contract deadline

    b. A payment due to flooding

    c. An expedient payment for improper work

    d. None of the above

24. Which of the following statements is true about partnerships?

    a. Each partner may cause the other to be jointly liable for any action incurred during normal business

    b. Each partner is personally liable for his individual actions only

    c. Each partner may sell his share of the partnership without the consent of the other partners

    d. A general partner may discharge himself from all liability by declaring himself to be a limited partner

25. Which of the following acts requires payment of prevailing wage rates and fringe benefits on federally assisted construction?

    a. Davis-Bacon Act

    b. Walsh-Healey Public Contracts Act

    c. Wage Garnishment Law

    d. Fair Labor Standards Act

Notes

Notes

USEFUL INFORMATION

The most common forms of business organizations include sole proprietorships, corporations, general partnerships, and limited partnerships. Business types are typically chosen with the advice of an attorney and/or accountant based on the degree of liability the owner is willing to accept. In other words, the trade off for less liability is more regulation and a greater tax burden.

Costs that relate to the construction activity itself but are not easily identifiable costs of a given project is referred to as project overhead. Company overhead, on the other hand, are costs that continue even when no work is underway. Examples of company overhead include: office salaries, office rent, company insurance, and accounting fees. Examples of project overhead include: a supervisor's salary, permit fees, bonds, and temporary toilets.

For a contract to be valid, there must be a mutual understanding, it must contain an offer, an acceptance and consideration. If the initial offer is not accepted, a counter offer is made. If the offer is accepted, it must be followed with consideration which is usually in the form of money. Many times negotiation precludes the offer. The difference between an offer and negotiation is that an offer is specific with price and completion dates. Mere negotiation is much less specific.

A bond is a three-party agreement between the insurance/surety company, the contractor and the owner (contractor's client). Upon issuance of the bond, the surety company assumes the liability for the contractor. The only party able to claim against a bond purchased by a contractor is the owner. Many different types of bonds exist such as: bid bond, performance bond, and payment bond. In the private sector of commercial construction, it is very common for the contractor to require the subcontractor to purchase a bond. The Miller Act requires bonds on federal construction projects.

Each employer with 11 or more employees is required to maintain records of all occupational injuries and illnesses regardless of their severity. Employers with 10 or fewer employees enjoy less stringent requirements by OSHA regarding reporting and record keeping; however, all employers must report fatalities and incidents resulting in multiple hospitalization. If a business changes ownership, an owner is responsible for recording and reporting work-related injuries and illnesses only for that period of the year during which he owned the establishment.

The Fair Labor Standards Act requires payment of federal minimum wage and overtime payment of at least one and one half the employee's regular rate of pay for hours worked over 40 during a fixed and recurring 168 consecutive hour workweek. Many times, supervisors and members of management, are exempt from many of these requirements. Employers should be careful to understand that a salaried employee does not always mean the FLSA rules of overtime will not apply.

To prove both employment eligibility and identity, a Form I-9 must be completed within 3 days of hiring a new employee. This form serves as documentation that the employer has obtained the necessary documents for verification. In the event a driver's license is provided to prove identity, a social security card must be provided to prove employment eligibility. The I-9 form must be kept on file for at least three years from the date of employment or for one year after the employee leaves the job, whichever is later.

There are two basic types of accounting: accrual basis and cash basis. Under the cash basis of accounting, income is recorded when it is received and records expenses when the payment is made. In contrast, under the accrual basis of accounting, income is recognized when it is earned regardless of when it is received and expenses are recorded when incurred regardless of when they are paid.

A qualifying agent is a full-time employee or officer of a company that has either volunteered or been selected to be the test-taker for the entity. Typically, the company's financials, rather than the qualifying agent, is presented to the appropriate agency for licensing consideration. When a qualifying agent terminates the relationship, the license will remain the property of the company. The company has a specified amount of time to replace the agent.

Notes

Notes

PART NINE
OSHA

The following section is based on OSHA 1926, Construction Industry Regulations. Generally, only a small portion of plumbing exams pertain to this subject matter. If the OSHA book is an approved reference material listed for your exam, these questions represent the types of questions that will be covered on your exam.

Notes

OSHA PRACTICE EXAM

1. The maximum slope for an 8' excavation in soil type B:

   a. ¾:1

   b. ½:1

   c. 1:1

   d. 2:1

2. The minimum perpendicular clearance between fixed ladder rungs, cleats, steps and any obstruction behind the ladder is:

   a. 6"

   b. 7"

   c. 8"

   d. 9"

3. The use of electrical cords for hoisting or lowering shall ______.

   a. Be allowed

   b. Not be allowed

   c. Only be permitted when load is less than 20 lbs.

   d. Only be permitted when load is less than 30 lbs.

4. When necessary to require employees to work in a trench, the excavated materials shall be placed a minimum of ____ feet away from the edge of the trench.

   a. 0

   b. 1

   c. 2

   d. 3

5. Eye protection near dangerous working conditions ______.

   a. Is required at the employee's expense
   b. Is required at the employer's expense
   c. Is not required, except by union regulations
   d. Is not required

6. Who is responsible for the fire safety and prevention on a jobsite through completion of the project?

   a. OSHA
   b. Employer
   c. Fire marshall
   d. Building inspector

7. When employees are required to be in trenches 4 feet deep or more, an adequate means of exit such as a ladder or steps shall be provided and located so as to require no more than ____ feet of lateral travel.

   a. 10
   b. 15
   c. 25
   d. 30

8. How much water is allowed in an excavation before work must cease:

   a. 3" or less at edge
   b. 2" or less at edge
   c. 1" or less at edge
   d. None

Notes

Notes

9. The minimum illumination in active general construction storage areas is ____.

   a. 3 foot candles

   b. 5 foot candles

   c. 10 foot candles

   d. 30 foot candles

10. After being sent to the job site, a first aid kit is checked at least ______ .

    a. Weekly

    b. Bi-weekly

    c. Monthly

    d. Quarterly

11. OSHA requires a ladder in every excavation in excess of:

    a. 5'

    b. 3'

    c. 4'

    d. 6'

12. Compressed air shall not be used for cleaning purposes, except where reduced to less than ________ psi.

    a. 10

    b. 20

    c. 30

    d. 70

13. Sand is considered what soil type?

    a. A

    b. B

    c. C

    d. D

14. According to OSHA, when employees are exposed to inhalation, ingestion, skin absorption, or contact with any substance at concentration above the _______ they shall be evacuated unless specific forms of personnel protective equipment or devices are provided.

    a. Threshold limit value
    b. PPM mandate
    c. PEL levels
    d. Oxygen deprivation level

15. According to OSHA the height at which no fall protection is required is ____ feet.

    a. 6
    b. 8
    c. 10
    d. 12

16. Circular saws shall be equipped with a _________ switch.

    a. position on-off
    b. momentary on-off
    c. lock on control
    d. constant pressure switch

17. Combustible debris should be removed from a construction job site _______________.

    a. when the job is complete
    b. only in the event of hazardous fire conditions
    c. periodically by any means
    d. at regular intervals

Notes

Notes

18. When using a stepladder:

    a. The top step can be used
    b. The cross-bracing on the back can be used as a step
    c. Both hands must be used on the sides
    d. The ladder should be inspected by a competent person for visible defects on a periodic basis

19. When employees are required to be in trenches 4 feet or more, an adequate means of egress shall be provided and located so as to require not more lateral travel than ______.

    a. 10'
    b. 12'
    c. 15'
    d. 25'

20. The maximum allowable slope for excavations less than 20 feet in depth for type C soil is ____________________.

    a. ¾ horizontal to 1 vertical
    b. 1 horizontal to 1 vertical
    c. 1½ horizontal to 1 vertical
    d. 2 horizontal to 1 vertical

21. A 10 foot deep trench is to be dug in granular sand. What is the max allowable bank slope?

    a. 1:1
    b. 1½:1
    c. ½:1
    d. ¼:1

22. When using a portable ladder for reaching an elevated surface or for ingress or egress of an excavation, the ladder must extend what minimum height above the surface?

a. 2.5'

b. 3'

c. 3.5'

d. 4'

23. ______ soil has a maximum allowable slope of 1½:1.

a. Type A

b. Type B

c. Type C

d. Type D

24. Each of the following is a soil type EXCEPT:

a. Gravel

b. Sediment

c. Sand

d. Clay

25. OSHA requires the employer to provide ______ for all employees/workers.

a. no smoking areas

b. a safe working environment

c. new tools

d. hardhats

Notes

Notes

PART TEN
Answer Keys

Notes

Part 2: Math Practice Exam Answer Key

1. b
2. a Travel = Advance or offset × 1.414; T = 28 × 1.414 = 39.592 ≈ 39½"
3. c offset = $\sqrt{r^2 + s^2}$; 0 = $\sqrt{8^2 + 20^2}$; 0 = $\sqrt{64^2 + 400^2}$; 0 = $\sqrt{464}$; 0 = 21.54;
   Travel = 21.54 × 1.414 = 30.45 ≈ 30½ inches
4. a Fall = length × % grade; F = 110' × 0.04 = 4.4 feet ≈ 4⅜ feet
5. d Pressure = 50'9" × 0.434; P = 50.75 × 0.434 = 22.03 psi
6. b Volume = 0.0408 × D^2 × Length; V = 0.0408 × 6^2 × 14;
   V = 0.0408 × 36 × 14 = 20.5632 gallons
7. c Head = Pressure × 2.3; H = 40 × 2.3 = 92 feet
8. d
9. b
10. d
11. e
12. a Fall = Length × Pitch F = 75 × ⅛ = 7⅝ = 9⅜ inches
13. d Pressure = Head × 0.434 Pressure = 30 × 0.434

    Pressure ≈ 13 psi loss (Uphill)

    Pressure = 60 − 13 = 47 psi
14. b
15. d
16. c Weight = 0.34 × D^2 × L; W = 0.34 × 10^2 × 10
    W – 0.34 × 100 × 10 = 340 pounds
17. b Volume = 0.7854 × D^2 × H; V = 0.7854 × 48^2 × 20;
    V = 0.7854 × 2304 × 20 = 36,191.232 cubic inches
18. b Volume = cubic inches ÷ 231; V = 36, 191.232 ÷ 231 = 156.672 gallons
19. c Travel = 1.414 × advance T = 1.414 × 48 = 67.872 inches ≈ 67⅞"
    67⅞" minus 3" (1½" for both fittings) = 64⅞" of pipe
20. a
21. d
22. b $4,450.00 × 60% = $2,670.00
23. c Payment = Base × Rate; Rate = Payment ÷ Base
    R = 900 ÷ 2195 = 0.41 = 41%
24. a
25. c

Notes

Part 3: Pipe, Fittings and Symbols Questionnaire Answer Key

1. d
2. b
3. b
4. d
5. a
6. c
7. e
8. d
9. b
10. b
11. a
12. b
13. a
14. c
15. a
16. b
17. c
18. d
19. d
20. c
21. a
22. b
23. d
24. a
25. d

Notes

Part 4: IPC Practice Exam #1 Answer Key

1. b 305.1
2. a 305.6
3. d 402.4
4. b Table 308.5
5. b 419.2
6. b 412.3
7. c Table 303.4
8. c Table 403.1 & 403.3
9. b 307.5
10. b 409.2
11. d 305.7
12. b Table 308.5
13. c 403.6
14. b 417.4
15. b 412.4
16. a 314.1
17. c 308.5
18. c Table 403.1
19. b 410.1
20. d 401.2
21. d 308.3
22. d 417.5
23. b 305.5
24. d 416.2
25. d 305.8

Notes

Part 4: IPC Practice Exam #2 Answer Key

1. d 602.3.1
2. d 504.7.2
3. c Table 604.3
4. a 603.2
5. b 606.5.4
6. b 608.3.1
7. a 501.6
8. d 504.4
9. c 602.3.5.1
10. d 605.23
11. d 605.11
12. c Table 608.1
13. a 608.8.1
14. b 502.3
15. d Table 604.5
16. a 505.1
17. b 603.1
18. d 605.20
19. b 606.2
20. d Table 604.4
21. c Table 605.3
22. d 605.23
23. c Table 606.5.4
24. c 608.6.1
25. a Table 606.5.7

Notes

Part 4: IPC Practice Exam #3 Answer Key

1. b 701.7
2. d 708.2
3. d 713.9 & definition Chapter 2
4. b 711.2
5. c 713.11.4
6. d Table 710.1(1)
7. a 705.16.3
8. b Table 706.3 footnote
9. c Table 704.1
10. b Table 702.2
11. a 705.19
12. a 708.3.3
13. c 712.2 exception
14. a Table 710.1(2)
15. b Table 712.4.2
16. a 712.3.2
17. c 705.14.2
18. b Table 709.1
19. b 708.3.2
20. d Table 709.2
21. c 708.7
22. d 709.3
23. a Table 709.1
24. a 708.8
25. b Table 709.1

Notes

Part 4: IPC Practice Exam #4 Answer Key

1. b Table 709.1
2. d Table 710.1(1) footnote: "Minimum size building drain serving a water closet shall be 3".
3. b Table 709.1 & Table 710.1(2) $2 + 2 + 2 = 6$ dfu
4. b Table 709.1 6 sleeping rooms × 6 dfu = 36

   Note: The question is not asking for total of all branch intervals.
5. a Table 710.1(2) 2000 dfu ÷ 5 floors = 400
6. c Table 710.1(1) 120 dfu × 2 branch intervals = 240 dfu
7. c Table 710.1(1)
8. b Table 710.1(1)

Notes

Part 4: IPC Practice Exam #5 Answer Key

1. a 802.2
2. d 802.4
3. d 903.2
4. d 916.3
5. c Table 916.5.1
6. a 908.3
7. a 917.4
8. b 905.3
9. c 803.2
10. b 901.2
11. b 911.1
12. d Table 916.1
13. a 908.1
14. b 916.4.1
15. a Table 906.1
16. b 803.1
17. c Table 909.3
18. d 914.2
19. c 917.3
20. d Table 909.3
21. d 802.2.1
22. a 904.6
23. c 902.2
24. c Table 906.1
25. a 916.1

Notes

Part 4: IPC Practice Exam #6 Answer Key

1. d 1002.4
2. a Table 1102.5
3. a 1106.3
4. c 12,000 sq. ft. ÷ 6 leaders = 2,000
5. d 1113.1.2
6. a 1113.14 exception
7. d 1002.3
8. d 1101.2
9. b 1109.1
10. d 1105.1
11. c 1002.1
12. a Table 1003.3.4.1
13. b Table 1106.2
14. c 1111.1
15. d 1113.1.4
16. a 1110.4
17. d 1111.1
18. b 1003.4.2.2
19. d 1002.2
20. a 1003.5
21. b 1103.3
22. c 1003.4.2.1
23. d Table 1106.3
24. b 1110.4
25. c 1106.3

Notes

Part 4: Storm Drain Diagram Answer Key

Leader Size:	5"
Horizontal A:	6"
Horizontal B:	8"
Horizontal C:	8"
Horizontal D:	10"

Notes

Part 4: Plumbing Isometrics Answer Key

	# Fixtures		dfu Value		Total
Water closets	8	×	6	=	48
Urinals	2	×	4	=	8
Lavatories	16	×	1	=	16
Floor drains	7	×	2	=	14
Shower heads	20	×	2	=	40
Water fountains	2	×	½	=	1
Washing machines	5	×	3	=	15
Laundry tray	1	×	2	=	2
			Total	=	144

Notes

Part 4: Plumbing Isometrics Answer Key (cont.)

Horizontal Drain Sizing (Figure 3)

Tip: When calculating a drainage fixture unit total for a horizontal drain, make a list of each type of fixture on the branch. This will help ensure all fixture types are accounted for. From this point, an accurate tally of drainage fixture units can be determined to correctly size the horizontal drain.

	Fixture Type	# of		dfu Value	Total
Horizontal Branch A:	Water Closet	3	×	6	18
	Floor Drains	2	×	2	4
	Urinals	2	×	4	8
					= 30

Horizontal branch A has a total of 30 dfus. Using Table 710.1(2) an accurate drain size can be determined.

What is the minimum drain size for horizontal branch A? ***Ans:*** *4"*

Use the example on page 89 to successfully complete the rest of the worksheet.

	Fixture Type	# of		dfu Value		Total
Horizontal Branch B:	Lavatories	8	×	1	=	8
	Floor Drains	1	×	2	=	2
	* Shower Dains	4	×	5	=	20
				Total	=	30

** Shower Drains = # of Shower heads ÷ # of floor drains in shower area*

Shower drains and floor drains have a different dfu value because shower drains are required to discharge a larger quantity of water.

	Fixture Type	# of		dfu Value		Total
Horizontal Branch B:	Shower Drains	4	×	5	=	20
	Fountains	2	×	½	=	1
	Floor Drains	1	×	2	=	2
	Laundry Tray	1	×	2	=	2
				Total	=	30

Notes

Part 4: Plumbing Isometrics Answer Key (cont.)

	Fixture Type	# of		dfu Value		Total
Horizontal Branch D:	Washing mach.	5	×	3	=	15
	Floor Drains	1	×	2	=	2
				Total	=	17
Horizontal Branch E:	Water Closets	5	×	6	=	30
	Floor Drains	1	×	2	=	2
				Total	=	32
Horizontal Branch F:	Lavatories	8	×	1	=	8
	Floor Drains	1	×	2	=	2
				Total	=	10

Each branch dfu total can now be used to size the horizontal drains.

1. What is the minimum drain size for horizontal branch B? 4"
2. What is the minimum drain size for horizontal branch C? 4"
3. What is the minimum drain size for horizontal branch D? 3"
4. What is the minimum drain size for horizontal branch E? 4"
5. What is the minimum drain size for horizontal branch F? 2½"

Sizing Building Drain (Figure 3)

When sizing a specific section of a building drain, one must first determine the total number of drainage fixture units flowing into that particular zone of pipe. Using Table 710.1(1), choose the correct column according to the given slope and size that specified section of drain pipe.

Building Drain A–B: This zone is conveying all fixture units discharged from horizontal branch A: (30 dfus). The slope, given in Figure 3, is ¼".

Column 4 of Table 710.1(1) determines the maximum number of dfus allowed to connect to a certain diameter of pipe having a ¼" slope.

Notes

Part 4: Plumbing Isometrics Answer Key (cont.)

	Total dfu Value		Building Drain Size
Building Drain A–B:	30	=	3"
Building Drain B–C:	60	=	4"
Building Drain C–D:	85	=	4"
Building Drain D–E:	102	=	4"
Building Drain E–F:	134	=	4"

Vent Sizing (Figure 3)

IPC 2003, Table 916.1: This is the most common table used for sizing vents. The drain size and the number of fixture units being vented must be known in order to properly size the vent.

As if you notice, in the first column, there are multiple listings of the same diameter of pipe. To ensure the correct diameter is chosen, the total fixture units being vented must be selected from column 2. The next step is to determine the developed length of vent pipe, which is usually given in the figure illustrated or the question. Everything needed to determine the vent size is in Table 916.1. Make sure you pay close attention to what each column represents.

	Drain Size	# FU Vented	Developed Length		Vent Size
Horizontal Branch A:	3"	30	30'	=	2"
Horizontal Branch B:	4"	30	30'	=	2"
Horizontal Branch C:	4"	25	20'	=	2"
Horizontal Branch D:	3"	17	30'	=	1½"
Horizontal Branch E:	4"	32	30'	=	2"
Horizontal Branch F:	2½"	10	30'	=	1½"

Notes

Part 4: Isometric Practice Exam Answer Key

Page 2: Top

1. d
2. d

Page 2: Bottom

1. a
2. a
3. c

Page 3:

1. d
2. a

Page 4:

1. c
2. b
3. d
4. c

Page 5:

1. b
2. a
3. c
4. b

Page 6:

1. c
2. b
3. d
4. a
5. d
6. c
7. b
8. c

Notes

Page 7:

1. b
2. d
3. a
4. c

Page 8:

5. d
6. b
7. c
8. a
9. d
10. b
11. c
12. a
13. d

Page 9:

14. c
15. b
16. d
17. b

Notes

Part 5: IPC 2003 Final Review Exam Answer Key

1. c Table 608.15.1
2. a 604.9
3. d 312.7
4. d 708.3.3 (must be greater than 45 degrees)
5. a 502.3
6. d 405.3.1
7. c Table 308.5
8. b Table 608.17.1
9. c Table 709.1
10. a 802.4
11. c Table 403.1
12. b 417.3
13. d 604.5
14. b Table 710.1(2)
15. b 603.1
16. c 1107.3
17. a 604.8
18. d Table 916.1
19. c 312.3
20. b 306.3
21. d 403.2 exception #2
22. b 405.3.1
23. a Table 709.1
24. a Table 909.3
25. d 302.1

Notes

Part 6: IFGC 2003 Practice Exam Answer Key

1. a 404.9.1
2. b Table 403.9.2
3. c Table 402.2
4. a 403.4.2
5. d 403.13
6. b 403.10.2
7. d 411.1.2
8. b Table 415.1
9. c 410.2
10. b 302.3.2
11. a 413.7
12. b Table 402.2 and Table 402.4(2)
13. a 402.6
14. c 302.3.3
15. c Chapter 2 definitions

Notes

Part 6: Gas Pipe Sizing Questionnaire Answer Key

1. d
2. c
3. c
4. d
5. c
6. d
7. c
8. b
9. b
10. b

Notes

Part 7: General Knowledge Practice Exam #1 Answer Key

1. d
2. c
3. d
4. b
5. a
6. c
7. d
8. c
9. d
10. b
11. c
12. c
13. a
14. a
15. c
16. c
17. b
18. c
19. c
20. a
21. b
22. d
23. a
24. b
25. b

Notes

Part 7: General Knowledge Practice Exam #2 Answer Key

1. d
2. e
3. b
4. e
5. c
6. a
7. d
8. d
9. c
10. b
11. d
12. c
13. b
14. c
15. b
16. a
17. a
18. d
19. d
20. b
21. c
22. a
23. a
24. c
25. b

Notes

Part 7: General Knowledge Practice Exam #3 Answer Key

1. b
2. a
3. c
4. b
5. b
6. b
7. c
8. c
9. a
10. b
11. b
12. d
13. c
14. a
15. b
16. d
17. a
18. c
19. c
20. a
21. d
22. b
23. b
24. a
25. a

Notes

Part 7: General Knowledge Practice Exam #4 Answer Key

1. c
2. b
3. a
4. d
5. a
6. c
7. d
8. a
9. c
10. d
11. b
12. b
13. a
14. b
15. d
16. b
17. d
18. a
19. c
20. d
21. a
22. a
23. b
24. a
25. c

Notes

Part 7: General Knowledge Practice Exam #5 Answer Key

1. d
2. d
3. c
4. a
5. d
6. d
7. c
8. b
9. b
10. d
11. a
12. b
13. b
14. d
15. d
16. c
17. d
18. d
19. a
20. d
21. c
22. c
23. a
24. a
25. c

Notes

Part 7: General Knowledge Practice Exam #6 Answer Key

1. b
2. c
3. a
4. c
5. d
6. a
7. c
8. b
9. d
10. d
11. b
12. a
13. a
14. b
15. c
16. d
17. a
18. a
19. b
20. b
21. b
22. d
23. c
24. b
25. a

Notes

Part 7: "Identify" Quiz Answer Key

1. a
2. b
3. c
4. c
5. b
6. d
7. d
8. d
9. d
10. b
11. c
12. d
13. a
14. b
15. d
16. c
17. a
18. a
19. c
20. b
21. d
22. c
23. a
24. c
25. a

Notes

Part 8: Project Management Practice Exam #1 Answer Key

1. a
2. c
3. c
4. a
5. c
6. d
7. c
8. a
9. c
10. c
11. d
12. a
13. c
14. d
15. b
16. d
17. d
18. b
19. d
20. c
21. a
22. a
23. a
24. a
25. a

Notes

Part 8: Project Management Practice Exam #2 Answer Key

1. b
2. a
3. b
4. c
5. a
6. b
7. c
8. a
9. b
10. a
11. b
12. a
13. b
14. b
15. c
16. c
17. a
18. b
19. b
20. b
21. b
22. d
23. a
24. a
25. a

Notes

Part 9: OSHA Practice Exam Answer Key

1. c 1926 Subpart P, Appendix B (Table B-1)
2. b 1926.1053 (a) #13
3. b 1926.302(a)(2)
4. c 1926.651(J)(2)
5. b 1926.102
6. b 1926.24
7. c 1926.651
8. d 1926.651(h)(1)
9. a 1926.56 (d)(3)
10. a 1926.50 (d)(2)
11. c 1926. 651(c)(2)
12. c 1926.302(b)(4)
13. c 1926 Subpart P Appendix A
14. a 1926.55 (a)
15. a 1926.501(b)(1)
16. d 1926.300(d)(3)
17. d 1926.25
18. d 1926.1053(b)
19. d 1926.651(j)(2)
20. c 1926 Subpart P Appendix A
21. b 1926 Subpart P Appendix A
22. b 1926.1053(b)(1)
23. c 1926 Subpart P Appendix A
24. b 1926 Subpart P Appendix A
25. d 1926.28

ABOUT THE AUTHOR

American Contractors Exam Services has successfully prepared candidates for state required licensing exams since 1991. A team of full time instructors comprises field experience, training expertise and formal education to formulate an effective and dynamic approach to concisely present pertinent test information in a seminar format. The seminars are conducted throughout the United States utilizing multimedia and interactive technology.

Utilizing effective graphics, pictures, and bulleted text, complex material is presented in an easy-to-understand format. This simplified approach has led to thousands of success stories and happy clients. One is sure to benefit from any material written or presented by American Contractors Exam Services.